B[illegible]UX

[illegible]IVES AND COACHING STOCK

SECOND EDITION

The complete guide to all Locomotives and Coaching Stock of SNCB/NMBS, NS and CFL

Brian Garvin & Peter Fox

PLATFORM 5

Published by Platform 5 Publishing Ltd., Lydgate House, Lydgate Lane, Sheffield S10 5FH, England.

Printed by Bayliss Printing Co. Ltd., Turner Road, Worksop, Notts.

ISBN 0 906579 96-1

Double-decker push-pull set, with Driving Open Second 50 84 26–37 107–8 leading, approaches Amsterdam Sloterdijk on 01/08/88 forming the 16.52 Amsterdam CS–Den Helder peak service. Class 1600 electric loco. 1633 is pushing at the rear.
David Brown

CONTENTS

INTRODUCTION

This book contains full details of all locomotives, multiple units and coaching stock of the state railways of the 'Benelux' countries, i.e. Belgium, the Netherlands and Luxembourg. Information is generally updated to Spring 1989.

Three years have passed quickly by since the first edition appeared. In that time many changes have taken place especially in Belgium and more are in the pipeline.

Belgium. Restructuring of the railway system continues with the former eight traffic areas having been reduced to five. The IC/IR services introduced in 1984 have settled down with slight modifications. After a large loss on passenger traffic, numbers are increasing again. Few changes have taken place in the EMU fleet but that is about to change. The old postal units have now been replaced by some new converted units and have entered full service with the summer 1989 timetable. New surburban units are being introduced which should see off some of the older units and now the "Break" units are to be made up to 3-car sets by the insertion of a new trailer. Continuing branch line closures have reduced the DMU fleet whilst freight traffic changes have decimated the main line diesel fleet with Classes 59 and 60 all but gone and the ranks of Class 80, 83 and 84 shunters reduced. The future however is good, as the Channel Tunnel and LGV Nord will bring more traffic.

Netherlands. Passenger traffic here is booming and more and more trains are needed. In the short term the NS is hiring loco-hauled coaches from West Germany and Belgium.In the long term a new batch of locos (Class 1700) and double-deck coaches will ease things. However more EMUs are on order. A 4 -car version of the IC3 unit is envisaged and some prototype local train EMUs are also due (SM90). On the diesel scene teething troubles with the new 6400s have meant a longer lease of life for the 2400s. Remote control of more diesel locos is envisaged in order to make freight operations more competitive.

Luxembourg. After many years of stability it is now all change on the CFL system as electrification heads north towards Gouvy. The section to Diekirch has already been completed but the new Class 2000 EMUs are not due until 1990. Meanwhile CFL is hiring some SNCF Z 6300s to help out. Acquisitions in the last few years have been mainly departmental shunters etc., reflecting the large amount of engineering work taking place. For the future some new electric locomotives are envisaged.

This expanded edition includes for the first time details of coaching stock in Belgium and Luxembourg. There is also a 'Museums and Museum Lines' section as in other recent titles in this series.

Good travelling!

GETTING THERE

By sea: Sealink offer a Harwich–Hoek van Holland (7 hour crossing) service; North Sea Ferries, Hull–Zeebrugge/Rotterdam (14 hrs); P & O, Dover–Zeebrugge/Oostende (4 hrs), and a less frequent Felixtowe–Zeebrugge (5½ hrs); Olau line, Sheerness–Vlissingen (7 hrs). The fastest sea crossing is by P & O jetfoil from Dover to Oostende (1½ hrs).

By air, the best fare deals and choice of flights are from all five London airports and 16 regional airports, direct to Amsterdam's modern rail-served airport at Schiphol. Air UK offer many flights and Dan Air and British Midland are very price-competitive. Other operators are British Airways, KLM, Aer Lingus, Netherlands BV, Transavia and NLM. One can also fly to Brussels from larger British airports. BA and Luxair provide direct Heathrow–Luxembourg flights.

AMENDMENTS

Please note the following amendments to the inside front cover of this book:

FSRE is Schaarbeek Electric.
LL & MKM. Correct spelling should read Arlon.
BACK COVER PHOTOGRAPHS: The Class 2700 is not dual voltage. Belgian locos run over NS metals to Maastricht under reduced power. There are now only three CFL Class 1600s, 1601 having finally been scrapped after being out of use for a number of years.

ROVER TICKETS

Railway enthusiasts will find railrover tickets very useful. At the time of writing these cost as follows: SNCB: £26 2nd class and £39 1st class for any five days out of seventeen (£20 or £30 age 6–25); and £46 2nd class and £69 1st class for sixteen days' consecutive travel. There is also a railcard (£8) which gives 50% reduction on both 1st and 2nd class tickets. NS: £21.50 2nd class and £32.50 1st class for three days, and £30 2nd class and £44 1st class for seven days. Benelux Tourrail Card (valid over the whole of the Benelux network): £37 2nd class and £56 1st class for any five days out of sixteen. There are reductions for children.

CLASS DIMENSIONS

Principal dimensions are given for each class in metric units. Standard abbreviations used in this book are:

km/h kilometres per hour
kN kilonewtons
kW kilowatts

ACKNOWLEDGEMENTS

We would like to thank all who have helped with the writing of this book, especially Messrs. P. Heppenstall and C. Appleby, Jan de Bruin, Messrs. M. Barthel and A. Ritz-Beckius at Luxembourg Depot, CFL, The public relations office of the SNCB/NMBS, also Messrs. G. Verhoeven (Département de Matériel) and L. Veriter (Salzinnes Works), the NS public relations office in London, also the Groupement Belge pour le Promotion et l'Exploitation Touristique du Transport Ferroviare (GTF), Liège and FEBELRAIL.

The following magazines have provided useful information.

La Vie du Rail, LCGB Bulletin. Op de Rails, Rail Hobby, Rail Kroniek, Rail Review, Rail Photo, Trans Fer.

VEHICLE TYPE CODES FOR MULTIPLE UNITS AND HAULED STOCK

These are given in the continental system with the British codes in parentheses for multiple units.

(1) CONTINENTAL SYSTEM:

f* Bicycle Van (Dutch – bicycles = fietsen)
m* Motor (with driving cab(s))
s* Driving Trailer
A 1st Class
B 2nd Class
D Luggage, i.e., vehicle with luggage space and guard's compartment
R Restaurant
K Buffet Kitchen
P Post, i.e., vehicle with compartment(s) for mail (and guard)

* NS only.

Examples:

BD Second Class with luggage/guard's compartment.
AB Composite

Note–The continental system does not differentiate between open and compartment stock or indicate toilet facilities.

(2) BRITISH SYSTEM:

Coaching Stock codes are as used in Multiple Unit Pocket Book and Coaching Stock Pocket Book, e.g., F first, S second, C composite, B brake, O open, K side corridor with lavatory, so semi-open.

The number of seats, lavatory compartments and washrooms are shown as nF nS nL nW, e.g.: 80S 2L 2W has 80 second class seats, two lavatory compartments and two washrooms.

UIC HAULED STOCK NUMBERING SYSTEM

Loco-hauled coaches are numbered according to the UIC standard system

The number consists of four pairs of digits, which describe owner, speed, heating type etc., a three digit serial number and a check digit. The system is as follows:

(a) Digits 1 and 2 indicate the exchange condition.

50 Passenger coach. Internal use only
51 Passenger coach. International use
60 Departmental passenger coach
61 Special service hauled stock
71 Sleeping cars. International use

(b) Digits 3 and 4. These give the railway of origin, e.g.:

80 DB (German Federal Railways)
82 CFL (Luxembourg Railways)
84 NS (Netherlands Railways)
87 SNCF (French Railways)
88 SNCB (Belgian Railways)

(c) Digit 5. The fifth digit gives the class or type of vehicle.

1 First class
2 Second class
3 Composite
4 Couchette–first class
5 Couchette–second class or composite
6 Sleeping car–first class
7 Sleeping car–second class or composite
8 Special purpose vehicle
9 Postal or luggage van

(d) Digit 6. The sixth digit gives the number of compartments (or windows/seating bays in the case of open stock).

0 10 compartments or bays
1 11 compartments or bays
2 12 compartments or bays
3 six-wheeled carriage
4 four-wheeled carriage
5 Reserved
6 6 compartments or bays or double decker
7 7 compartments or bays
8 8 compartments or bays
9 9 compartments or bays

(d) Digits 7 and 8. These give the maximum speed and type of operation.

Digit 7:

0 120 km/h, electric heating
1 120 km/h, dual heating
2 120 km/h, steam heating (except for 29–no heating)
3 121–140 km/h, electric heating
4/5 121–140 km/h, dual heating
6 121–140 km/h, steam heating (except for 69–no heating)
7 141–160 km/h, electric heating
8 141–160 km/h, dual heating (except for 84–steam, 89–no heating)
9 Above 160 km/h

Digit 8 sometimes depends on digit 7, but in general:

0 All voltages
6 All voltages except 3000 V dc
7 1500 V dc or 50 Hz ac
8 3000 V dc
9 If digit 7 is 0,1,7–3000 V dc. If digit 7 is 3,4,5–3000 V dc + 1000 V 16 2/3 Hz ac.

Digits 9, 10 and 11. These give the serial number of the individual vehicle.

Digit 12. This gives the check digit. Multiply as follows:

Digit	1 2	3 4	5 6	7 8	9 10	11
x	2 1	2 1	2 1	2 1	2 1	2

Add all resultant digits and subtract the last number of the result from 10. This gives the check digit, example:

	5 0	8 4	2 9	3 7	2 4 8	
x	2 1	2 1	2 1	2 1	2 1 2	
	1+0+0 +	1+6+4 +	4+9 +	6+7 +	4+4+1+6	= 53

Subtract 3 from 10—check digit is 7.

BUILDERS

The following builder codes are used in this publication:

ABR	Ateliers Belges Réunis SA.
ACEC	SA Ateliers de Constructions Electriques de Charleroi.
AFB	Société Anglo-Franco-Belge des Ateliers de la Croyère.
Allan	N.V. Allan & Co.
Alsthom	Société Générale de Constructions Electriques et Mechaniques Alsthom.
BBC	Brown Boveri.
BL	Brissonneau & Lotz.
BLC	Usines de Braine le Comte.
BM	Baume et Marpent SA.
BN	La Brugeoise et Nivelles SA.
BND	La Brugeoise, Nicaise & Delcuve SA.
BR	British Railways.
Beijnes	Koninklijke Fabriek van Rijtuigen en Spoorwagens JJ Beijnes NV.
CEM	Companie Electro Mecanique.
CF	Etablissements Carel Fouché & Cie.
CFCF	Constructions Ferroviaires du Centre à Familleureux.
CWFM	SNCB Central Workshops, Mechelen.
Cockerill	SA Cockerill-Ougrée.
Deutz	Klöckner, Humboldt, Deutz AG, Köln.
Donelli	Donelli SpA, Poviglio, Reggio, Emilia, Italy.
Duewag	Duewag Uerdingen.
Dyle	SA des Ateliers de la Dyle.
Energie	SA Energie Marcinelle, Charleroi.
FUF	Forges, Usines et Fonderies, Haine St. Pierre.
Fam.	Ateliers de Contruction de Familleureux, SA.
Fiat	Fiat. Italy.
GM	General Motors, USA.
Geismar	Société des Anciens Etablissements L. Geismar, Colmar, France.
Germain	Ateliers Germain, Monceau.
Heemaf	Heemaf NV.
Henschel	Henschel und Sohn Gmbh, Kassel.
Jung	Arnold Jung Lokomotivfabrik Gmbh, Kirchen an der Sieg.
Krupp	Freid. Krupp, Essen.
MTE	Société de Matériel de Traction Electrique, France.
Marelli	SA Ercolle Marelli, Milano.
MV	Metropolitan Vickers, Gorton, Manchester.
Niv.	SA Les Ateliers Metallurgiques de Nivelles.
Oerlikon	Oerlikon, Switzerland.
Rag.	SA Usines Ragheno.
SEM	Société d'Electricité et de Méchanique, Charleroi.
SEMG	Société d'Electricité et de Méchanique, Gent.
SIG	SIG, Neuhausen, Switzerland.
Schneider	Société des Forges et Ateliers du Creusot, Usines Schneider.
Smit	Smit, Slikkerveer.
Talbot	Waggonfabrik Talbot, Aachen.
Werkspoor	Werkspoor NV, Utrecht.
Westwaggon	Vereinigte Westdeutsche Waggonfabriken AG.

1. BELGIAN NATIONAL RAILWAYS (SNCB/NMBS)

GENERAL

Belgium is a two language country and has presented your authors with problems over which name/terminology to use. The north of the country is Flemish speaking (Flemings) whilst the south is French speaking (Walloons). Brussels is an exception being the only official dual-language area. We have compromised by using the terminology normally used in the various parts of the country but beg forgiveness for any errors. An exception is Brussels where the English version is used in preference to Brussel (Flemish) or Bruxelles (French).

Because of the language problem, Belgian National Railways uses the logo (B) rather than initials as these differ according to the language used! The Flemish use NMBS (Nationale Maatschappij der Belgische Spoorwegen) whilst the French use SNCB (Société Nationale des Chemins de Fer Belges).

NUMBERING SYSTEM

The SNCB list is quite straight forward. The present scheme dates from 01/01/71 and is as follows:

0001–1000	Electric multiple units. (Leading '0' not carried).
1001–2000	Electric locomotives. High speed range and multi-voltage.
2001–3000	Electric locomotives. Mixed traffic.
3001–4000	Not used.
4001–5000	Diesel railcars.
5001–6000	Diesel locomotives. Higher power.
6001–7000	Diesel locomotives. Medium power.
7001–8000	Diesel shunting locomotives. Heavy duty.
8001–9000	Diesel shunting locomotives. Medium power.
9001–9999	Diesel shunting locomotives. Low power.

ORGANISATION, DEPOTS & STABLING POINTS

Since the first edition of this book appeared, there have been many reorganisations on Belgian Railways. These have eventually caught up with the traffic and motive power sections which have been reduced from eight areas to five regions. The list of stabling points is far from complete and does not include every station where the odd EMU may be stabled.

Central Region.

Depots: Brussels Midi/Zuid, Schaarbeek Diesel, Schaarbeek Electric.
Departmentals: Forest Vorst, Schaarbeek.
Stabling points: Brussels Midi/Zuid station, Ottignies.

North East Region.

Depots: Antwerpen Dam, Hasselt.
Departmentals: Antwerpen Oost, Leuven.
Stabling points: Aarschot, Antwerpen Centraal, Antwerpen Noord, Antwerpen Schijnpoort, Berchem, Leuven Station, Leuven Yard, Mechelen, Mol, Turnhout.

North West Region.

Depots: Kortrijk, Merelbeke, Oostende
Departmentals: Brugge. (also maintains some FSD shunters).
Stabling points: Aalst, Brugge, Denderleeuw, Dendermonde, De Panne, Gent St. Pieters, Geraardsbergen, Oudenaarde.

South East Region.

Depots: Bertrix, Latour, Gouvy, Kinkempois (Liège), Ronet (Namur), Stockem.
Departmentals: Angleur, Arlon, Jemelle, Kinkempois, Visé.
Stabling points: Arlon, Huy, Jemelle, Liège Guillemins, Liers, Namur, Stockem Yard, Virton, Welkenraedt.

South West Region.

Depots: Haine St. Pierre, Monceau (Charleroi), St. Ghislain.
Departmentals: Charleroi Sud, Mons.
Stabling points: Ath, Charleroi Sud, Châtelet, Mons, Tournai.

The SNCB has used codes for depots for many years dating back to the days of the telegraphic system. These codes are still in use today as official abbreviations. However on locomotives the allocation is normally stencilled on in full on the mainframe somewhere below the cab. EMUs and diesel railcars do not normally carry their allocations but somtimes the code will be found on them against repair data etc.

LIVERIES

The Belgian Railways have had a number of liveries in recent years. The old standard livery was green with a narrow yellow stripe for both electric and diesel locomotives, with a wider stripe for diesel shunters. The class 25.5 and the former Benelux EMUs were painted dark blue with a yellow stripe.

The main-line diesel locomotive fleet was being repainted in green with a much broader yellow stripe, but this then gave way to the current scheme of yellow with a green stripe. No distinction is made in this book between the two liveries. An exception is electric train-heat fitted diesels, which are painted blue with a yellow stripe. All new electric locomotives with the exception of class 11 are now being painted blue and yellow and this livery is now being applied to older electric locos. Class 11 are painted in the new Benelux livery which reflects their international use, as the upper half of the body is painted Bordeaux red, the new SNCB Inter-City colour, whilst the lower part is painted yellow, the NS colour.

Diesel railcars are painted yellow and red, whilst old EMUs are painted plain green. New EMUs are painted Bordeaux red, and certain old ones are being repainted in that colour, as are the type AM75/76/77 four-car units, which were originally painted grey and orange.

Coaching stock liveries are dealt with later.
Note: Where a dot is shown in the livery column, the livery carried is not known.

WORKSHOPS

There are only two workshops for the general overhaul of locomotives and multiple units. However it should be stated that most of the main depots can undertake quite heavy repairs. The main works are:

Mechelen: This works deals with all EMUs, diesel railcars and electric locomotives of classes 15, 16, 22, 23, 25, 25.5 and 28.
Salzinnes: This works deals with all the diesel locomotives and electric locomotives of classes 11, 12, 18, 20, 21, 26 and 27.

PLACE NAMES

Listed below are some of the towns and cities with their alternative rendering. Lille in France is included since at Antwerpen you will not see Lille quoted on the departure board! Fl. Flemish, Fr. French.

Usual Name	*Alternative*	*Usual Name*	*Alternative*
Aalst	Alost (Fr.)	Kortrijk	Courtrai (Fr.)
Arlon	Aarlen (Fl.)	Leuven	Louvain (Fr.)
Antwerpen	Anvers (Fr.)	Lille	Rijssel (Fl.)
Ath	Aat (Fl.)	Liège	Luik (Fl.)
Brugge	Bruges (Fr.)	Mechelen	Malines (Fr.)
Dendermonde	Termonde (Fr.)	Mons	Bergen (Fl.)
Gent	Gand (Fr.)	Namur	Namen (Fl.)
Geraardsbergen	Grammont (Fr.)	Oudenarde	Audenarde (Fr.)
Ieper	Yprès (Fr.)	Tournai	Doornik (Fl.)

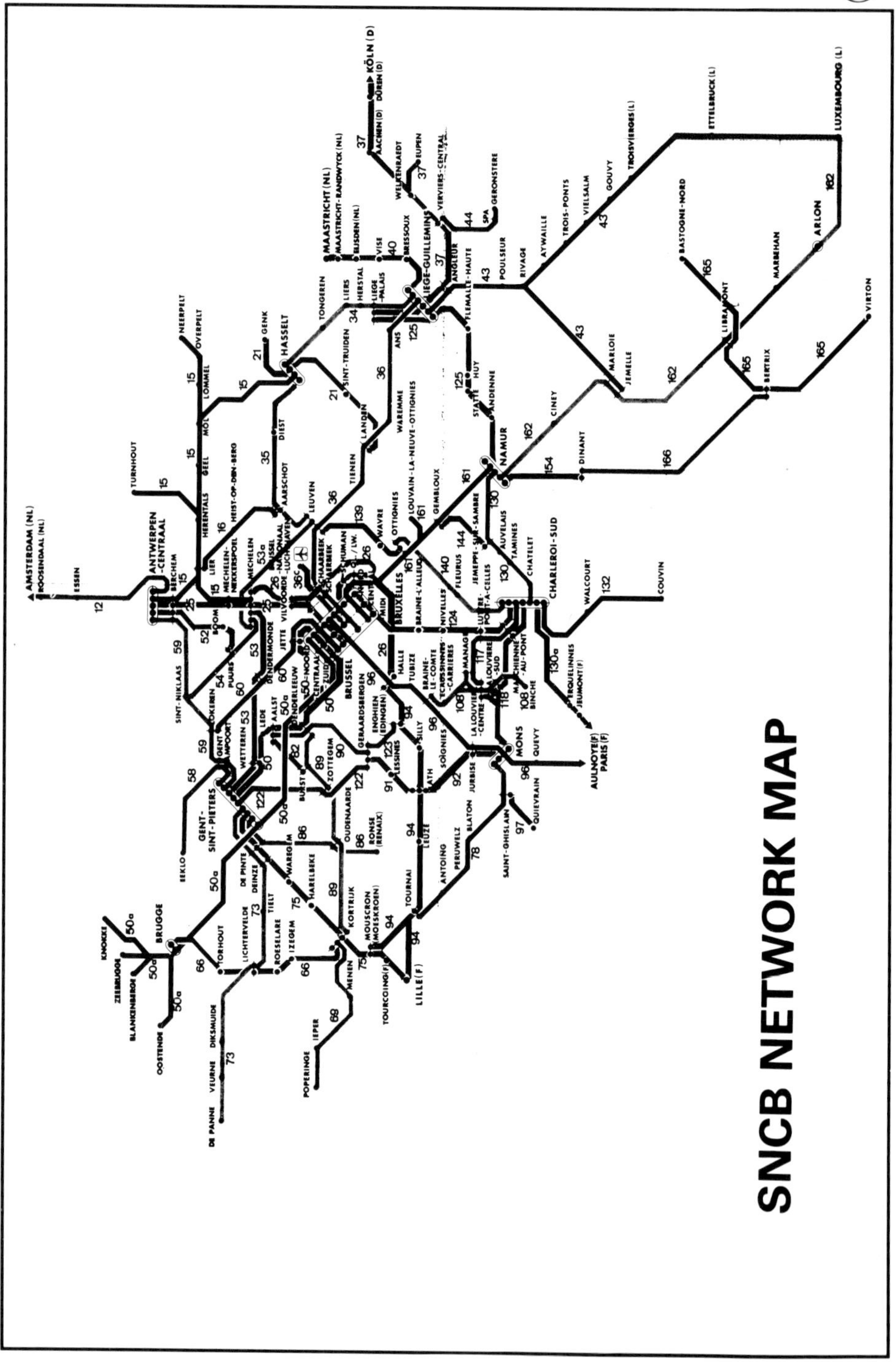
SNCB NETWORK MAP
AMSTERDAM (NL)
ROOSENDAAL (NL)
ESSEN
ANTWERPEN-CENTRAAL
BERCHEM
TURNHOUT
HERENTALS
GEEL
MOL
LOMMEL
OVERPELT
NEERPELT
GENK
HASSELT
TONGEREN
MAASTRICHT (NL)
VISE
LIERS
HERSTAL
LIEGE-PALAIS
BRESSOUX
LIEGE-GUILLEMINS
ANGLEUR
FLEMALLE-HAUTE
WELKENRAEDT
EUPEN
AACHEN (D)
KÖLN (D)
DÜREN (D)
VERVIERS-CENTRAL
SPA
GERONSTERE
POULSEUR
RIVAGE
AYWAILLE
TROIS-PONTS
VIELSALM
GOUVY
TROISVIERGES (L)
ETTELBRUCK (L)
LUXEMBOURG (L)
ARLON
MARBEHAN
LIBRAMONT
BASTOGNE-NORD
BERTRIX
VIRTON
MARLOIE
JEMELLE
CINEY
DINANT
NAMUR
ANDENNE
STATTE
HUY
ANS
SINT-TRUIDEN
LANDEN
WAREMME
TIENEN
LEUVEN
AARSCHOT
DIEST
LIER
MECHELEN
HEIST-OP-DEN-BERG
VILVOORDE
JETTE
BRUSSEL
BRUXELLES
MIDI
OTTIGNIES
WAVRE
LOUVAIN-LA-NEUVE-OTTIGNIES
GEMBLOUX
FLEURUS
JEMEPPE-SUR-SAMBRE
AUVELAIS
TAMINES
CHATELET
CHARLEROI-SUD
WALCOURT
COUVIN
ERQUELINNES
JEUMONT (F)
AULNOYE (F)
PARIS (F)
NIVELLES
BRAINE-L'ALLEUD
BRAINE-LE-COMTE
ECAUSSINNES-CARRIERES
MANAGE
LA LOUVIERE-CENTRE
LA LOUVIERE-SUD
BINCHE
MONS
QUEVY
QUIEVRAIN
SAINT-GHISLAIN
JURBISE
SOIGNIES
HALLE
TUBIZE
ENGHIEN (EDINGEN)
SILLY
ATH
LESSINES
GERAARDSBERGEN
ZOTTEGEM
OUDENAARDE
RONSE (RENAIX)
BLATON
PERUWELZ
ANTOING
TOURNAI
LEUZE
LILLE (F)
MOUSCRON (MOESKROEN)
TOURCOING (F)
KORTRIJK
MENEN
HARELBEKE
WAREGEM
DEINZE
DE PINTE
IEPER
POPERINGE
IZEGEM
ROESELARE
TIELT
LICHTERVELDE
TORHOUT
DIKSMUIDE
VEURNE
DE PANNE
OOSTENDE
BRUGGE
KNOKKE
ZEEBRUGGE
BLANKENBERGE
EEKLO
GENT-SINT-PIETERS
GENT-DAMPOORT
WETTEREN
LEDE
AALST
DENDERLEEUW
DENDERMONDE
LOKEREN
SINT-NIKLAAS
PUURS
BOOM

ELECTRIC MULTIPLE UNITS

SNCB EMUs operate in fixed formations and therefore only unit numbers are quoted. All classes of EMU may work in multiple with one another except for types AM80/82/83, and type AM86, which may work with other members of the same type only. "Type" refers to the year in which the batch of units were ordered. Seating is 2+2 in first class and 2+3 in second class.

TYPE AM50/53 — 2 CAR UNITS

These units are restricted to the Roosendaal–Antwerpen–Brussels–Charleroi route (lines 12, 25, 27 and 124) where there are high platforms. 001–034 are type AM50 and 035–049 are type AM53.

AB+BD (DMCOL–DMBSOL).

Wheel Arrangement: A1–1A + A1–1A.
Built: 1950.
Builder-Mech. Parts: BND.
Builder-Elec. Parts: ACEC, SEMG.
Traction Motors: 4 x 155 kW.
Accommodation: 28F 5S 1L + 78S 1L.
Weight: 47 + 45 tonnes.
Length over couplings: 22.201 + 22.201 m.
Max. Speed: 130 km/h.

Originally numbered 228.010–034.

010 FSRE
011 FSRE
013 FSRE
014 FSRE
015 FSRE
016 FSRE
017 FSRE
018 FSRE
019 FSRE
020 FSRE
022 FSRE
023 FSRE
024 FSRE
025 FSRE
026 FSRE
027 FSRE
029 FSRE
030 FSRE
031 FSRE
032 FSRE
033 FSRE
034 FSRE
035 FSRE
036 FSRE
037 FSRE
038 FSRE
039 FSRE
040 FSRE
041 FSRE
042 FSRE
043 FSRE
044 FSRE
045 * FSRE
046 * FSRE
047 * FSRE
048 * FSRE
049 * FSRE

TYPE AM62-3 — 2 CAR UNIT

This unit was rebuilt in 1987 from unit 151 and is fitted with three-phase asynchronous motors and prototype suburban BN bogies incorporating pneumatic suspension. It is now a high platform unit, and works over the same lines as types AM50/53. Details as for type AM62 except the traction motors which are 4 x 200 kW.

Originally numbered 228.151.

050 FSRE

TYPE AM54 — 2 CAR UNITS

NK units are used on lines 21, 25, 34–37, 44, 50A, 89, 125, FEO units are used on lines 26, 130,139, 140, 144, 161 and 161D, whilst the two GT units work with type AM55.

AB+BD (DMCOL–DMBSOL).

Wheel Arrangement: A1–1A + A1–1A.
Built: 1954.
Builder-Mech. Parts: 051–082 BN, 083–100 Rag, 101–128 Fam (AB), Germain (BD).
Builder-Elec. Parts: ACEC, SEMG.
Traction Motors: 4 X 155 kW.
Accommodation: 28F 58S 1L + 85S 1L
Weight: 43 + 42 tonnes.
Length over couplings: 22.640 + 22.640 m.
Max. Speed: 130 km/h.

Originally 050–128. 055 was destroyed in an accident 04/10/67 but reinstated in 1968 with the original AB coach and BD ex 526. 062 was destroyed by fire at Quévy in 1970 and was withdrawn with 050 being renumbered to 062. Originally numbered 228.050–128.

051	NK	067	NK	090	FEO	107	FEO
052	NK	068	NK	092	NK	108	FEO
053	NK	069	NK	094	GT	112	FEO
054	NK	071	NK	095	GT	113	FEO
055	NK	072	NK	096	FEO	114	FEO
056	NK	073	NK	097	FEO	115	FEO
057	NK	075	NK	098	FEO	116	FEO
058	NK	076	NK	099	FEO	119	FEO
059	NK	077	NK	100	FEO	121	FEO
060	NK	078	NK	101	FEO	122	FEO
061	NK	079	NK	102	FEO	124	FEO
063	NK	081	NK	103	FEO	125	FEO
064	NK	082	NK	104	FEO	126	FEO
065	NK	083	NK	105	FEO	127	FEO
066	NK	088	NK	106	FEO		

TYPE AM56 — 2 CAR UNITS

Post-war american influence is reflected in these units as they are finished off with Budd stainless steel panelling. These units are used on services on lines 26, 130, 139, 140, 144, 161 and 161D.

AB+BD (DMCOL–DMBSOL).

Wheel Arrangement: A1–1A + A1–1A.
Built: 1956.
Builder-Mech. Parts: BND (under licence from Budd, Philadelphia, PA, USA).
Builder-Elec. Parts: ACEC, SEMG.
Traction Motors: 4 x 155 kW.
Accommodation: 28F 56S 1L + 84S 1L.
Weight: 40 + 39 tonnes.
Length over couplings: 22.640 + 22.640 m.
Max. Speed: 130 km/h.

Originally numbered 228.129–150.

129	FEO	135	FEO	141	FEO	146	FEO
130	FEO	136	FEO	142	FEO	147	FEO
131	FEO	137	FEO	143	FEO	148	FEO
132	FEO	138	FEO	144	FEO	149	FEO
133	FEO	139	FEO	145	FEO	150	FEO
134	FEO	140	FEO				

TYPE AM62/63/65 — 2 CAR UNITS

These units are in general use on local services over a wide area. FKR units are used on lines 50, 53, 60, 89 and 90, FHS units on lines 15, 16, 35, 36, 52 and 59, FSRE units on lines 12, 25, 26 and 124 and LK units on lines 50, 51A, 51B, 66 and 89.

ABD+B (DMBCO–DMSO).

Wheel Arrangement: A1–1A + A1–1A.
Built: 1962–1965. 151–210 are type AM62, 211–250 are type AM63 and 251–270 are type AM65.
Builder-Mech. Parts: BN, Rag, BLC, ABR, Germain, CWFM.
Builder-Elec. Parts: ACEC.
Traction Motors: 4 x 155 kW.
Accommodation: 28F 48S + 104S.
Weight: 49 +50 tonnes.
Length over couplings: 23.287 + 23.287 m.
Max. Speed: 130 km/h.

Electro-pneumatic brakes. Disc brakes. Originally numbered 228.153–270.

153	FKR	158	FKR	163	FKR	168	FKR
154	FKR	159	FKR	164	FKR	169	FKR
155	FKR	160	FKR	165	FKR	170	FKR
156	FKR	161	FKR	166**R**	FKR	171	FKR
157	FKR	162	FKR	167	FKR	172	FKR

173 FKR
174 FKR
175 FKR
176 FKR
177 FKR
178 FKR
179 FKR
180 FKR
181 FKR
182 FKR
183 FKR
184 FKR
185 FKR
186 FKR
187 FKR
188 FKR
189 FKR
190 FKR
191 FKR
192 FKR
193 FKR
194 FKR
195 FKR
196 FKR
197 FKR
198 FKR
199 FKR
200 FKR
201 FKR
202 FKR
203 FKR
204 FKR
205 FKR
206 FKR
207 FHS
208 FHS
209 FHS
210 FHS
211 FHS
212 FHS
213 FHS
214 FHS
215 FHS
216 FHS
217 FHS
218 FHS
219 FHS
220 FHS
221 FHS
222 FHS
223 FHS
224 FHS
225 FHS
226 FHS
227 FHS
228 FHS
229 FHS
230 FHS
231 FSRE
232 FSRE
233 FSRE
234 FSRE
235 FSRE
236 FSRE
237 FSRE
238 FSRE
239 FSRE
240 FSRE
241 FSRE
242 FSRE
243 LK
244 LK
245 LK
246 LK
247 LK
248**R** LK
249**R** LK
250 LK
251 LK
252 LK
253 LK
254 LK
255 LK
256 LK
257 LK
258 LK
259 LK
260 LK
261 LK
262 LK
263 LK
264 LK
265 LK
266 LK
267 LK
268 LK
269 LK
270 LK

TYPE AM80/82/83 2 CAR UNITS

These are the latest generation of Belgian EMUs and feature thyristor control. They have very modern interiors similar to M4 loco-hauled stock. Nicknamed "break" units (french for "estate car"). Automatic doors have an audible warning of closure. Pressure ventilation. MKM units are used on routes 161 and 162, NK units on routes 34, 75A, 78, 97, 112, 118, 125 and 130 and FSD units on routes 21, 21A, 36, 40, 50, 51, 51B and 161. Seventy new trailers have been ordered to make units up to 3 car sets.

B+ABD (DMSOL–PDTBCOL).

Wheel Arrangement: Bo–Bo + 2–2.
Built: 1980–84. 301–335 are type AM80, 336–370 are type AM82 and 371–440 are type AM83.
Builder-Mech. Parts: BN.
Builder-Elec. Parts: ACEC.
Traction Motors: 4 x 310 kW.
Accommodation: 99S 1L + 32F 40S 1L
Weight: 60 + 47 tonnes.
Length over couplings: 25.425 + 25.425 m.
Max. Speed: 160 km/h.

Disc brakes. Electro-pneumatic and regenerative braking.
f–Modified Faiveley pantograph.
b–Brecknell-Willis pantograph.

301**R** MKM
302**R** MKM
303**R** MKM
304**R** MKM
305**R** MKM
306**R** MKM
307**R** MKM
308**R** MKM
309**R** MKM
310**R** MKM
311**R** MKM
312**R** MKM
313**R** MKM
314**R** MKM
315**R** MKM
316**R** MKM
317**R** MKM
318**R** MKM
319**R** MKM
320**R** MKM
321**R** MKM
322**R** MKM
323**R** MKM
324**R** MKM
325**R** MKM
326**R** MKM
327**R** MKM
328**R** MKM
329**R** MKM
330**R** MKM
331**R** MKM
332**R** MKM
333**R** MKM
334**R** MKM
335**R** MKM
336**R** MKM
337**R** NK
338**R** NK
339**R** NK
340**R** NK
341**R** NK
342**R** NK
343**R** NK
344**R** NK
345**R** NK
346**R** NK
347**R** NK
348**R** NK
349**R** NK
350**R** NK
351**R** NK
352**R** NK
353**R** NK
354**R** NK
355**R** NK
356**R** NK
357**R** FSD
358**R** FSD
359**R** FSD
360**R** FSD
361**R** FSD
362**R** FSD
363**R** FSD
364**R** FSD

365**R** FSD
366**R** FSD
367**R** FSD
368**R** FSD
369**R** FSD
370**R** FSD
371**R** FSD
372**R** FSD
373**R** FSD
374**R** FSD
375**R** FSD
376**R** FSD
377**R** FSD
378**R** f FSD
379**R** FSD
380**R** FSD
381**R** FSD
382**R** FSD
383**R** FSD
384**R** FSD
385**R** FSD
386**R** FSD
387**R** FSD
388**R** FSD
389**R** FSD
390**R** FSD
391**R** FSD
392**R** FSD
393**R** FSD
394**R** FSD
395**R** FSD
396**R** FSD
397**R** FSD
398**R** FSD
399**R** FSD
400**R** FSD
401**R** FSD
402**R** FSD
403**R** FSD
404**R** FSD
405**R** FSD
406**R** FSD
407**R** FSD
408**R** FSD
409**R** FSD
410**R** FSD
411**R** FSD
412**R** FSD
413**R** FSD
414**R** FSD
415**R** FSD
416**R** FSD
417**R** FSD
418**R** FSD
419**R** FSD
420**R** FSD
421**R** FSD
422**R** FSD
423**R** FSD
424**R** FSD
425**R** FSD
426**R** FSD
427**R** FSD
428**R** FSD
429**R** FSD
430**R** FSD
431**R** FSD
432**R** FSD
433**R** FSD
434**R** FSD
435**R** FSD
436**R** FSD
437**R** bFSD
438**R** bFSD
439**R** bFSD
440**R** bFSD

TYPE AM55 — 2 CAR UNITS

Used on local services. MKM units are used on lines 161 and 162 and GT units on lines 90, 96, 112, 117, 118, 124 and 130A.

ABD+B(DMBCOL–DMSOL).

Wheel Arrangement: A1–1A + A1–1A.
Built: 1955.
Builder-Mech. Parts: Dyle, AFB, FUF.
Builder-Elec. Parts: ACEC, SEMG.
Traction Motors: 4 x 155 kW.
Accommodation: 28F 39S 1L + 104S 1L.
Weight: 42 + 43 tonnes.
Length over couplings: 22.840 + 22.840 m.
Max. Speed: 135 km/h.

526 was withdrawn in 1957 after an accident in Brussels. One coach went to 055. 524/5/35/6 are reported to be for early withdrawal. Originally numbered 228.502–539.

502 MKM
503 MKM
504 MKM
505 MKM
506 MKM
508 MKM
509 MKM
510 MKM
511 MKM
512 MKM
514 MKM
515 MKM
516 MKM
517 MKM
518 GT
519 GT
520 GT
521 GT
522 GT
523 GT
524 GT
525 GT
527 GT
528 GT
529 GT
530 GT
531 GT
532 GT
534 GT
535 GT
536 GT
537 GT
538 GT
539 GT

TYPE AM70A (SABENA) — 2-CAR UNITS

Exclusively used on Brussels Airport services for which they were built until the train service alterations of 1984 since when they have also turned up on Brussels area local services in rush hours! Known as 'Sabena' or airport units they were originally numbered 851–856 and were renumbered to allow for the possible building of more four-car sets. A new livery is being considered. Lines 36 and 36C.

ABD+B (DMBCOL–DMSOL).

Wheel Arrangement: A1–1A + A1–1A.
Built: 1970.
Builder-Mech. Parts: BN, Rag, BLC, ABR.
Builder-Elec. Parts: ACEC.
Traction Motors: 4 x 170 kW.
Accommodation: 32F 20S + 99S 1L.
Total Weight: 52 + 52 tonnes.
Length over couplings: 22.840 + 22.840 m.

Max. Speed: 140 km/h.
Special Livery: Blue with white band.

Electro-pneumatic brakes. Disc brakes.

595**S**	FSRE	597**S**	FSRE	599**S**	FSRE	600**S**	FSRE
596**S**	FSRE	598**S**	FSRE				

TYPE AM66/70 — 2 CAR UNITS

GT units are used on lines 25, 96, 108, 116, 117 and 124 and LK units on lines 13, 15, 27, 29, 50, 51A, 51B, 53, 66, 69, 75 and 89. 656 was withdrawn in 1975.

ABD+B (DMBCO–DMSOL).

Wheel Arrangement: A1–1A + A1–1A.
Built: 1966/70.
Builder-Mech. Parts: BN, Rag, BLC, ABR.
Builder-Elec. Parts: ACEC.
Traction Motors: 4 x 170 kW.
Accommodation: 28F 48S + 104S 1L.
Weight: 52 + 56 tonnes.
Length over couplings: 23.287 + 23.287 m.
Max. Speed: 140 km/h.

Electro-pneumatic brakes. Disc brakes.
*–Fitted with Timken roller bearings.
Originally numbered 228.601–664.

601**R**	GT	617**R**	GT	633	* GT	649**R**	* LK
602**R**	GT	618**R**	GT	634	* GT	650	* LK
603**R**	GT	619	GT	635	* GT	651	* LK
604**R**	GT	620**R**	GT	636**R**	* GT	652	* LK
605	GT	621**R**	GT	637**R**	* LK	653	* LK
606	GT	622	GT	638	* LK	654	* LK
607**R**	GT	623	GT	639	* LK	655	* LK
608	GT	624	GT	640	* LK	657	* LK
609**R**	GT	625**R**	GT	641	* LK	658	* LK
610**R**	GT	626	GT	642	* LK	659	* LK
611	GT	627**R**	GT	643	* LK	660**R**	* LK
612**R**	GT	628	GT	644	* LK	661	* LK
613**R**	GT	629	GT	645**R**	* LK	662	* LK
614**R**	GT	630**R**	GT	646	* LK	663	* LK
615	GT	631**R**	* GT	647	* LK	664	* LK

TYPE AM70Th — 2 CAR UNITS

The 'Th' in the type classification denotes thyristor control, these being the first Belgian units so fitted. Timken roller bearings. Used on lines 34, 78, 97, 116, 117, 125 and 130.

ABD+B (DMBCO–DMSOL).

Wheel Arrangement: A1–1A + A1–1A.
Built: 1971.
Builder-Mech. Parts: CWFM.
Builder-Elec. Parts: ACEC.
Traction Motors: 4 x 170 kW.
Accommodation: 28F 48S + 104S 1L.
Weight: 53 + 56 tonnes.
Length over couplings: 23.287 + 23.287 m.
Max. Speed: 140 km/h.

Electro-pneumatic brakes. Disc brakes.

665	NK	668	NK	671**R**	NK	674**R**	NK
666	NK	669	NK	672	NK	675	NK
667	NK	670	NK	673	NK	676	NK

TYPE AM73/74/78/79 — 2 CAR UNITS

These are the production series of thyristor-controlled units and are used as follows: NK units on lines 32, 34, 36, 44, 78, 97, 116, 117, 118, 125 and 130. FSRE units on lines 12, 25, 26 and 124. FHS units on lines 15, 16, 21, 25, 35, 36, 94, 96, 117 and 123.

ABD+B (DMCO–DMSOL).

Wheel Arrangement: A1–1A + A1–1A.
Built: 1973–80.
Builder-Mech. Parts: BN, CFCF.
Builder-Elec. Parts: ACEC.
Traction Motors: 4 x 170 kW.
Accommodation: 28F 48S + 102S 1L.
Weight: 52 + 56 tonnes.
Length over couplings: 23.308 + 23.308 m.
Max. Speed: 140 km/h.

Electro-pneumatic brakes. Disc brakes.

677 NK
678 NK
679**R** NK
680 NK
681 NK
682 NK
683 NK
684 NK
685 NK
686 NK
687 NK
688 NK
689 NK
690 NK
691 NK
692 NK
693 NK
694 NK
695 NK
696 NK
697**R** NK
698 FSRE
699**R** FSRE
700 FSRE
701 FSRE
702 FSRE
703**R** FSRE
704**R** FSRE
705**R** FSRE
706 FSRE
707**R** FSRE
708**R** FSRE
709**R** FSRE
710 FSRE
711 FSRE
712**R** FHS
713 FHS
714**R** FHS
715 FHS
716**R** FHS
717**R** FHS
718 FHS
719 FHS
720 FHS
721 FHS
722**R** FHS
723**R** FHS
724**R** FHS
725**R** FHS
726**R** FHS
727**R** FHS
728 FHS
729**R** FHS
730 FHS
731 FHS
732 FHS
733 FHS
734 FHS
735 FHS
736 FHS
737 FHS
738 FHS
739 FHS
740 FHS
741 FHS
742 FHS
743 FHS
744 FHS
745 FHS
746 FHS
747 FHS
748 FHS
749 FHS
750 FHS
751 FHS
752 FHS
753 FHS
754 FHS
755 FHS
756 FHS
757 FHS
758 FHS
759 FHS
760 FHS
761 FHS
762 FHS
763 FHS
764 FHS
765 FHS
766 FHS
767 FHS
768 FHS
769**R** FHS
770 FHS
771 FHS
772 FHS
773 FHS
774 FHS
775 FHS
776 FHS
777 FHS
778 FHS
779 FHS
780 FHS
781 FHS
782 FHS

TYPE AM75/76/77 — 4 CAR UNITS

These thyristor-controlled units are gangwayed within the sets only. Pantographs are fitted to only one of the motor coaches. Pressure ventilation. Used on lines 12, 25, 50, 57, 59, 60, 96, 97 and 124.

AD+B+B+B (DTBFOL–PMSOL–MSOL–DTSOL).

Wheel Arrangement: 2–2 + Bo–Bo + Bo–Bo + 2–2.
Built: 1975–1979.
Builder-Mech. Parts: BN.
Builder-Elec. Parts: ACEC.
Traction Motors: 8 x 170 kW.
Accommodation: 56F 1L + 100S 1L + 106S 1L + 96S 1L.
Weight: 51 + 60 + 60 + 49 tonnes.
Length over couplings: 99.028 m.
Max. Speed: 140 km/h.

Electro-pneumatic brakes. Disc brakes.

801**R**	FSRE	812**R**	FSRE	823**O**	FSRE	834**O**	FSRE
802**R**	FSRE	813**R**	FSRE	824**R**	FSRE	835**O**	FSRE
803**R**	FSRE	814**R**	FSRE	825**R**	FSRE	836**O**	FSRE
804**R**	FSRE	815**R**	FSRE	826**R**	FSRE	837**O**	FSRE
805**R**	FSRE	816**R**	FSRE	827**R**	FSRE	838**O**	FSRE
806**R**	FSRE	817**R**	FSRE	828**O**	FSRE	839**O**	FSRE
807**R**	FSRE	818**R**	FSRE	829**R**	FSRE	840**O**	FSRE
808**R**	FSRE	819**R**	FSRE	830**R**	FSRE	841**O**	FSRE
809**R**	FSRE	820**R**	FSRE	831**R**	FSRE	842**O**	FSRE
810**R**	FSRE	821**R**	FSRE	832**O**	FSRE	843**O**	FSRE
811**O**	FSRE	822**R**	FSRE	833**O**	FSRE	844**O**	FSRE

TYPE AM86 — 2 CAR UNITS

The latest in Belgian EMUs, these are designed to replace some of the oldest in traffic of types AM50/53. They are the first Belgian EMUs to feature 2+2 seating in second class. Another innovation is the use of polyester sides and front nose which are stuck onto the main body. They are designed for eventual one-person operation and rear-view mirrors are fitted which are flush with the side of the vehicle when not in use. They are officially known as Sprinters, but their unusual front end appearance has led to them being nicknamed "Snorkels". Used on lines 25, 27, 52, 59 and 124.

AB+B (MSOL–DTCO).
Wheel Arrangement: Bo–Bo + 2–2.
Built: 1988–89.
Builder-Mech. Parts: BN.
Builder-Elec. Parts: ACEC.
Traction Motors: 4 x 172 kW type AE121N.
Accommodation: 16F 72S 1L + 86S.
Weight: 59 + 47 tonnes.
Length over couplings: . + . m.
Max. Speed: 120 km/h.
Electro-pneumatic brakes. Disc brakes.

901**R**	FSRE	910**R**	FSRE	919**R**	FSRE	928**R**	FSRE
902**R**	FSRE	911**R**	FSRE	920**R**	FSRE	929**R**	FSRE
903**R**	FSRE	912**R**	FSRE	921**R**	FSRE	930**R**	FSRE
904**R**	FSRE	913**R**	FSRE	922**R**	FSRE	931**R**	FSRE
905**R**	FSRE	914**R**	FSRE	923**R**	FSRE	932**R**	FSRE
906**R**	FSRE	915**R**	FSRE	924**R**	FSRE	933**R**	FSRE
907**R**	FSRE	916**R**	FSRE	925**R**	FSRE	934**R**	FSRE
908**R**	FSRE	917**R**	FSRE	926**R**	FSRE	935**R**	FSRE
909**R**	FSRE	918**R**	FSRE	927**R**	FSRE		

TYPE AM54P — 2 CAR UNITS

These units were converted at Mechelen Works from type AM54 during 1988. They have replaced six older units converted from type 25 stock. These latest units have been designed to handle mail carried in containers which will be loaded into the coaches and have roller shutter doors. A total of 86 containers can be carried. Full details of their use has not been received but their main sphere of activity is expected to be Gent–Antwerpen, Gent–Brussels–Liège and Antwerpen–Brussels.

D+D (2DMBP).
Wheel Arrangement: A1–1A + A1–1A.
Built: 1954.
Builder-Mech. Parts: See type AM54.
Builder-Elec. Parts: ACEC, SEMG.
Traction Motors: 4 x 155 kW.
Weight: 43 + 42 tonnes.
Length over couplings: 22.640 + 22.640 m.
Max. Speed: 130 km/h.

961 (080)	**P**	NK	965 (117)	**P**	NK	969 (091)	**P**	NK	973 (120)	**P**	NK
962 (085)	**P**	NK	966 (123)	**P**	NK	970 (093)	**P**	NK	974 (110)	**P**	NK
963 (109)	**P**	NK	967 (118)	**P**	NK	971 (128)	**P**	NK	975 (111)	**P**	NK
964 (086)	**P**	NK	968 (074)	**P**	NK	972 (084)	**P**	NK			

ELECTRIC LOCOMOTIVES

CLASS 11 Bo–Bo

These dual voltage (1500/3000 V dc) locos are a development of class 27 but lower powered. They are used on the revamped Brussels–Amsterdam service. The SNCB provides the locos for these push & pull Inter-City services and the NS provides the coaching stock. Originally planned to be numbered 1101–1112, the higher numbers were eventually decided on to avoid conflicting with the NS 1100 class. Otherwise the class is part of the new family of locos of classes 11, 12, 21 and 27. Thyristor control.

Built: 1985–6.
Builder-Mech. Parts: BN.
Builder-Elec. Parts: ACEC.
Traction Motors: 4 x LE622S frame mounted.
One Hour Rating: 3310 kW. **Total Weight**: 85 tonnes.
Maximum Tractive Effort: 234 kN. **Length over Buffers**: 18.65 m.
Driving Wheel Dia.: 1250 mm. **Max. Speed**: 160 km/h.

Rheostatic braking.

1181 **B**	FBMZ	1184 **B**	FBMZ	1187 **B**	FBMZ	1190 **B**	FBMZ
1182 **B**	FBMZ	1185 **B**	FBMZ	1188 **B**	FBMZ	1191 **B**	FBMZ
1183 **B**	FBMZ	1186 **B**	FBMZ	1189 **B**	FBMZ	1192 **B**	FBMZ

CLASS 12 Bo–Bo

Introduced in 1986, this latest member of the new generation of electric locos is another dual voltage machine (3000 V dc/25 kV ac) and is used on the Antwerpen–Mouscron–Lille services. They are also passed for working to Paris if required. Thyristor control.

Built: 1986.
Builder-Mech. Parts: BN.
Builder-Elec. Parts: ACEC.
Traction Motors: 4 x LE622S frame mounted.
One Hour Rating: 3310 kW. **Total Weight**: 85 tonnes.
Maximum Tractive Effort: 234 kN. **Length over Buffers**: 18.65 m.
Driving Wheel Dia.: 1250 mm. **Max. Speed**: 160 km/h.

Electro-pneumatic braking. Rheostatic braking.

1201 **N**	FBMZ	1204 **N**	FBMZ	1207 **N**	FBMZ	1210 **N**	FBMZ
1202 **N**	FBMZ	1205 **N**	FBMZ	1208 **N**	FBMZ	1211 **N**	FBMZ
1203 **N**	FBMZ	1206 **N**	FBMZ	1209 **N**	FBMZ	1212 **N**	FBMZ

CLASS 15 Bo–Bo

These are triple voltage locos (1500/3000 V dc, 25 kV ac) for through workings to the NS and SNCF systems. The class is used on IC/TEE services Paris–Brussels–Amsterdam and Paris–Liège, but from 1988 have been banned from the NS. Class 25.5 being used instead between Brussels and Amsterdam. The pantograph at No. 1 end is for dc and that at No. 2 end is for ac.

Built: 1962.
Builder-Mech. Parts: BN.
Builder-Elec. Parts: ACEC.
Traction Motors: 4 x ES541 frame mounted.
One Hour Rating: 2780 kW. **Total Weight**: 77.7 tonnes.
Maximum Tractive Effort: 170 kN. **Length over Buffers**: 17.75 m.
Driving Wheel Dia.: 1250 mm. **Max. Speed**: 160 km/h.

Electro-pneumatic brakes. Originally numbered 150.001–003/011/012.

1501 **N**	FBMZ	1503 **N**	FBMZ	1504 **N**	FBMZ	1505 **N**	FBMZ
1502 **N**	FBMZ						

CLASS 16 Bo–Bo

When introduced these locos brought a bold new styling to the SNCB and an electric blue livery instead of the dark green then prevailing. These are true international locos. and work on 1500/3000 V dc, 15/25 kV ac. The main sphere of operation is Paris–Brussels/Liège/ Köln and Oostende–Brussels–Köln. For a few years one member used to work through to Basel on a holiday train. Note that these locos now rarely work through into the Netherlands. No. 1 end has two pantographs, one for 15 kV ac and one for 1500/3000 V dc whilst No. 2 end has the 25 kV ac pantograph.

Built: 1966.
Builder-Mech. Parts: BN.
Builder-Elec. Parts: ACEC.
Traction Motors: 4 x ES541 frame mounted.
One Hour Rating: 2780 kW. **Total Weight**: 82.6 tonnes.
Maximum Tractive Effort: 196 kN. **Length over Buffers**: 16.65 m.
Driving Wheel Dia.: 1250 mm. **Max. Speed**: 160 km/h.

Electro-pneumatic brakes. Originally numbered 160.001–004/021–024.

1601 **N** FBMZ
1602 **N** FBMZ
1603 **N** FBMZ
1604 **N** FBMZ
1605 **N** FBMZ
1606 **N** FBMZ
1607 **N** FBMZ
1608 **N** FBMZ

CLASS 18 C–C

These four voltage locos are in effect a more powerful version of the SNCF CC 40100 and feature the monomotor bogies which the French find so popular. The pantograph layout is No. 1 end 1500/3000 V dc, No. 2 end two pantographs, one for 15 kV ac and one for 25 kV ac. Duties are similar to class 16.

Built: 1973.
Builder-Mech. Parts: BN.
Builder-Elec. Parts: Alsthom.
Traction Motors: 2 x TDQ662C1 frame mounted.
One Hour Rating: 4450 kW. **Total Weight**: 113 tonnes.
Maximum Tractive Effort: 196 kN. **Length over Buffers**: 22.06 m.
Driving Wheel Dia.: 1100 mm. **Max. Speed**: 180 km/h.
Special livery: These locos are blue and grey with a yellow stripe and stainless steel panelling.

Electro-pneumatic braking. Rheostatic braking.

1801 **S** NK
1802 **S** NK
1803 **S** NK
1804 **S** NK
1805 **S** NK
1806 **S** NK

CLASS 20 Co–Co

These thyristor controlled locos have had a chequered career with varying defects over the years which now appear to have been ironed out. The main use is on the Oostende–Brussels–Luxembourg artery as well as Namur–Antwerpen.

Built: 1975–77.
Builder-Mech. Parts: BN.
Builder-Elec. Parts: ACEC.
Traction Motors: 6 x LE772G frame mounted.
One Hour Rating: 5150 kW. **Total Weight**: 110 tonnes.
Maximum Tractive Effort: 314 kN. **Length over Buffers**: 19.50 m.
Driving Wheel Dia.: 1250 mm. **Max. Speed**: 160 km/h.

Electro-pneumatic brakes. Separately excited rheostatic brakes.

2001 **N** FEO
2002 FEO
2003 FEO
2004 **N** FEO
2005 FEO
2006 **N** FEO
2007 **N** FEO
2008 **N** FEO
2009 **N** FEO
2010 FEO
2011 FEO
2012 **N** FEO
2013 FEO
2014 FEO
2015 **N** FEO
2016 **N** FEO
2017 FEO
2018 FEO
2019 FEO
2020 FEO
2021 FEO
2022 FEO
2023 FEO
2024 FEO
2025 **N** FEO

CLASS 21 Bo–Bo

These are similar to class 27 but lower powered. Used on push-pull trains with M4 or M5 stock and freights.

Built: 1984–1987.
Builder-Mech. Parts: BN.

▲Class 16 quadruple voltage electric loco. No. 1604 at Leuven with the 12.34 Oostende-Köln formed mainly of I10 stock, but with a DB corridor second next to the loco. *David Brown*

▼Green-liveried class 20 No. 2004 at the head of the Iris at Brussels Midi on 18/08/86. This train is now a Eurocity train and runs from Oostende to Chur. The vehicle behind the loco is a coal-heated Wagons-Lits restaurant car dating from 1940–43. Such a vehicle is still included in the formation of this train. *David Brown*

Builder-Elec. Parts: ACEC.
Traction Motors: 4 x LE622S frame mounted.
One Hour Rating: 3310 kW. **Total Weight**: 85 tonnes.
Maximum Tractive Effort: 234 kN. **Length over Buffers**: 18.65 m.
Driving Wheel Dia.: 1250 mm. **Max. Speed**: 160 km/h.
Special livery: 2109 is in advertising livery.

Rheostatic brakes.

2101	**N**	FKR	2116	**N**	FKR	2131	**N**	FKR	2146	**N**	FKR
2102	**N**	FKR	2117	**N**	FKR	2132	**N**	FKR	2147	**N**	FKR
2103	**N**	FKR	2118	**N**	FKR	2133	**N**	FKR	2148	**N**	FKR
2104	**N**	FKR	2119	**N**	FKR	2134	**N**	FKR	2149	**N**	FKR
2105	**N**	FKR	2120	**N**	FKR	2135	**N**	FKR	2150	**N**	FKR
2106	**N**	FKR	2121	**N**	FKR	2136	**N**	FKR	2151	**N**	FKR
2107	**N**	FKR	2122	**N**	FKR	2137	**N**	FKR	2152	**N**	FKR
2108	**N**	FKR	2123	**N**	FKR	2138	**N**	FKR	2153	**N**	FKR
2109	**N**	FKR	2124	**N**	FKR	2139	**N**	FKR	2154	**N**	FKR
2110	**N**	FKR	2125	**N**	FKR	2140	**N**	FKR	2155	**N**	FKR
2111	**N**	FKR	2126	**N**	FKR	2141	**N**	FKR	2156	**N**	FKR
2112	**N**	FKR	2127	**N**	FKR	2142	**N**	FKR	2157	**N**	FKR
2113	**N**	FKR	2128	**N**	FKR	2143	**N**	FKR	2158	**N**	FKR
2114	**N**	FKR	2129	**N**	FKR	2144	**N**	FKR	2159	**N**	FKR
2115	**N**	FKR	2130	**N**	FKR	2145	**N**	FKR	2160	**N**	FKR

CLASS 22 Bo–Bo

A general purpose locomotive found all over the network. 2239–2250 were originally dual voltage (1500/3000 V dc) but are now believed to be standard with the first batch.

Built: 1954.
Builder-Mech. Parts: BN.
Builder-Elec. Parts: SEMG/ACEC.
Traction Motors: 4 x CF729 axle-hung.
One Hour Rating: 1880 kW. **Total Weight**: 87 tonnes.
Maximum Tractive Effort: 196 kN. **Length over Buffers**: 18.00 m.
Driving Wheel Dia.: 1262 mm. **Max. Speed**: 130 km/h.

Originally numbered 122.001–038/201/212.

2201	.	FGH	2214	**N**	FGH	2227	**N**	FKR	2239	.	FKR
2202	.	FGH	2215	**N**	FGH	2228	.	FKR	2240	**N**	FKR
2203	**N**	FGH	2216	.	FGH	2229	.	FKR	2241	**N**	FKR
2204	**N**	FGH	2217	**Y**	FGH	2230	**N**	FKR	2242	**N**	FKR
2205	.	FGH	2218	.	FGH	2231	.	FKR	2243	**N**	FKR
2206	**Y**	FGH	2219	.	FGH	2232	.	FKR	2244		FKR
2207	**Y**	FGH	2220	**Y**	FGH	2233	**N**	FKR	2245	.	FKR
2208	**N**	FGH	2221	.	FGH	2234	**N**	FKR	2246	**N**	FKR
2209	**Y**	FGH	2222	**Y**	FGH	2235	**N**	FKR	2247	**N**	FKR
2210	**Y**	FGH	2223	**N**	FGH	2236	**N**	FKR	2248	**N**	FKR
2211	**Y**	FGH	2224	.	FGH	2237	**N**	FKR	2249	**N**	FKR
2212	**N**	FGH	2225	**N**	FGH	2238		FKR	2250	**N**	FKR
2213	**N**	FGH	2226	.	FKR						

CLASS 23 Bo–Bo

Another mixed traffic loco at work all over the system. Fitted with regenerative brakes for freight work and can work in multiple with others of the same class and class 26.

Built: 1955.
Builder-Mech. Parts: Niv.
Builder-Elec. Parts: ACEC/SEMG.
Traction Motors: 4 x CF729 axle-hung.
One Hour Rating: 1880 kW. **Total Weight**: 93.3 tonnes.
Maximum Tractive Effort: 196 kN. **Length over Buffers**: 18.00 m.
Driving Wheel Dia.: 1262 mm. **Max. Speed**: 130 km/h.

Regenerative braking.

Originally numbered 123.001–083.
2383 is fitted with special equipment for banking at Liège. It was renumbered 124.001 and then 2401 before becoming 2383.
† Has side portholes on bottom of bodyside and ventilation grills as on class 27.

2301	.	FEO	2322	**Y**	FEO	2343	.	FEO	2364	**N**	NK
2302	**N**	† FEO	2323	**Y**	FEO	2344	**N**	FEO	2365	.	NK
2303		FEO	2324	**N**	FEO	2345	**N**	FEO	2366	.	NK
2304	.	FEO	2325	.	FEO	2346		FEO	2367	**Y**	NK
2305	**N**	FEO	2326	**N**	FEO	2347		FEO	2368	**N**	NK
2306	**N**	FEO	2327	**N**	FEO	2348		FEO	2369	**N**	NK
2307	.	FEO	2328	**N**	FEO	2349	**N**	FEO	2370	**Y**	NK
2308	**Y**	FEO	2329	**N**	FEO	2350	.	FEO	2371	**N**	NK
2309	**N**	FEO	2330	**N**	FEO	2351		FEO	2372	**N**	NK
2310		FEO	2331	**N**	FEO	2352		FEO	2373	**Y**	NK
2311	.	FEO	2332	.	FEO	2353	**N**	FEO	2374	**Y**	NK
2312	.	FEO	2333	**N**	FEO	2354	.	FEO	2375	**N**	NK
2313	**N**	FEO	2334	**N**	FEO	2355	.	FEO	2376	**Y**	NK
2314	**N**	FEO	2335		FEO	2356	**N**	FEO	2377	**Y**	NK
2315	**N**	FEO	2336		FEO	2357	**N**	FEO	2378	**Y**	NK
2316	**N**	FEO	2337	**N**	FEO	2358		FEO	2379	**Y**	NK
2317	**N**	FEO	2338		FEO	2359	.	FEO	2380	**Y**	NK
2318	.	FEO	2339	**N**	FEO	2360	**N**	FEO	2381	**Y**	NK
2319	**N**	FEO	2340	**N**	FEO	2361	**N**	FEO	2382	.	NK
2320	**N**	FEO	2341	**N**	FEO	2362	**N**	FEO	2383	**N**	* NK
2321	**N**	FEO	2342	**N**	FEO	2363	**N**	FEO			

CLASS 25 Bo–Bo

This class is used on push & pull trains around Antwerpen using M2 stock. 2504 was rebuilt from 2557 (2521) in 1979.

Built: 1960–61.
Builder-Mech. Parts: BN.
Builder-Elec. Parts: ACEC/SEM.
Traction Motors: 4 x CF729 axle-hung.
One Hour Rating: 1880 kW.
Maximum Tractive Effort: 196 kN.
Driving Wheel Dia.: 1262 mm.
Total Weight: 84 tonnes.
Length over Buffers: 18.00 m.
Max. Speed: 130 km/h.

Originally numbered 125.001–014.

2501	**Y**	FBMZ	2505	.	FBMZ	2509	**Y**	FBMZ	2512	.	FBMZ
2502	.	FBMZ	2506	.	FBMZ	2510	.	FBMZ	2513	**Y**	FBMZ
2503	.	FBMZ	2507	.	FBMZ	2511	**Y**	FBMZ	2514	**Y**	FBMZ
2504	**N**	FBMZ	2508	**Y**	FBMZ						

CLASS 25.5 Bo–Bo

Formerly numbered 2515–22, these locos were modified in 1973 to dual voltage 1500/3000 V dc for working the push & pull Brussels–Amsterdam service for which they received an additional headlamp and a special blue livery. Note that only one pantograph is fitted. These locos have been replaced on these duties by class 11 but still work to Amsterdam on former cl. 15 duties. They are not compatible with the new Benelux push-pull stock. After an accident in 1979, 2557 and 2504 changed identities.

Built: 1960–61.
Builder-Mech. Parts: BN.
Builder-Elec. Parts: ACEC/SEM.
Traction Motors: 4 x CF729 axle-hung.
One Hour Rating: 1880 kW.
Maximum Tractive Effort: 196 kN.
Driving Wheel Dia.: 1262 mm.
Total Weight: 85 tonnes.
Length over Buffers: 18.00 m.
Max. Speed: 130 km/h.

Originally numbered 125.015–016, 140.001–006 (later 125.101–106).

2551 (2515)	**X**	FBMZ	2554 (2518)	**X**	FBMZ	2557 (2521)	**X**	FBMZ
2552 (2516)	**X**	FBMZ	2555 (2519)	**X**	FBMZ	2558 (2522)	**X**	FBMZ
2553 (2517)	**X**	FBMZ	2556 (2520)	**X**	FBMZ			

CLASS 26 B–B

These locos feature monomotor bogies by Schneider and have two gear ratios giving maximum speeds of 100/130 km/h. Seen all over Belgium, the class is fitted for multiple working with other members of the class and with class 23 for heavy freights.

Built: 1964–71.
Builder-Mech. Parts: BN.
Builder-Elec. Parts: ACEC.
Traction Motors: 2 x 2ES508 frame mounted.
One Hour Rating: 2580 kW. **Total Weight**: 82.4 tonnes.
Maximum Tractive Effort: 235 kN. **Length over Buffers**: 17.25 m.
Driving Wheel Dia.: 1150 mm. **Max. Speed**: 130 km/h.

2601–20 were formerly originally 126.001–005, 126.101–130.

2601	.	MKM	2610	.	MKM	2619	**N**	MKM	2628	**N**	MKM
2602	**N**	MKM	2611	**Y**	MKM	2620	**N**	MKM	2629	**N**	MKM
2603	**N**	MKM	2612	.	MKM	2621	**N**	MKM	2630	**N**	MKM
2604	**N**	MKM	2613	**N**	MKM	2622	**N**	MKM	2631	**N**	MKM
2605	.	MKM	2614	**N**	MKM	2623	**Y**	MKM	2632	**N**	MKM
2606	**N**	MKM	2615	**N**	MKM	2624	**N**	MKM	2633	**N**	MKM
2607	**N**	MKM	2616	**N**	MKM	2625	**Y**	MKM	2634	.	MKM
2608	.	MKM	2617	**Y**	MKM	2626	**N**	MKM	2635	.	MKM
2609	**N**	MKM	2618	**Y**	MKM	2627	**Y**	MKM			

CLASS 27 Bo–Bo

These are the first of the 1980s generation of electric locomotives and heralded a new era, being more powerful than their predecessors. Chopper control. Flexicoil suspension. They are often found working push & pull trains of the latest M4 IC stock.

Built: 1981–1983.
Builder-Mech. Parts: BN.
Builder-Elec. Parts: ACEC.
Traction Motors: 4 x LE921S frame mounted.
One Hour Rating: 4250 kW. **Total Weight**: 85 tonnes.
Maximum Tractive Effort: 234 kN. **Length over Buffers**: 18.65 m.
Driving Wheel Dia.: 1250 mm. **Max. Speed**: 160 km/h.

Electro-pneumatic brakes. Rheostatic brakes.

2701	**N**	FSD	2716	**N**	FSD	2731	**N**	NK	2746	**N**	NK
2702	**N**	FSD	2717	**N**	FSD	2732	**N**	NK	2747	**N**	NK
2703	**N**	FSD	2718	**N**	FSD	2733	**N**	NK	2748	**N**	NK
2704	**N**	FSD	2719	**N**	FSD	2734	**N**	NK	2749	**N**	NK
2705	**N**	FSD	2720	**N**	FSD	2735	**N**	NK	2750	**N**	NK
2706	**N**	FSD	2721	**N**	NK	2736	**N**	NK	2751	**N**	NK
2707	**N**	FSD	2722	**N**	NK	2737	**N**	NK	2752	**N**	NK
2708	**N**	FSD	2723	**N**	NK	2738	**N**	NK	2753	**N**	NK
2709	**N**	FSD	2724	**N**	NK	2739	**N**	NK	2754	**N**	NK
2710	**N**	FSD	2725	**N**	NK	2740	**N**	NK	2755	**N**	NK
2711	**N**	FSD	2726	**N**	NK	2741	**N**	NK	2756	**N**	NK
2712	**N**	FSD	2727	**N**	NK	2742	**N**	NK	2757	**N**	NK
2713	**N**	FSD	2728	**N**	NK	2743	**N**	NK	2758	**N**	NK
2714	**N**	FSD	2729	**N**	NK	2744	**N**	NK	2759	**N**	NK
2715	**N**	FSD	2730	**N**	NK	2745	**N**	NK	2760	**N**	NK

CLASS 28 Bo–Bo

Formerly numbered 2001–2003, these locos became class 28 in 1973 and with the recent withdrawal of class 29 they are now the oldest electric locos at work on the SNCB and are due for early withdrawal. They are relegated to light duties and can usually be found on pilot/e.c.s. duties at Brussels and Schaerbeek/Mechelen.
Built: 1949.
Builder-Mech. Parts: BM.
Builder-Elec. Parts: ACEC/SEMG.
Traction Motors: 4 x CF729 axle-hung.

One Hour Rating: 1985 kW.
Maximum Tractive Effort: 196 kN.
Driving Wheel Dia.: 1262 mm.

Total Weight: 85 tonnes.
Length over Buffers: 17.18 m.
Max. Speed: 130 km/h.

Originally numbered 120.001–003.

2801	.	FBMZ	2802	.	FBMZ	2803	**N**	FBMZ

▲Former 'Benelux' Class 25/5 dual-voltage (1500/3000 V dc) electric loco. No. 2555 passes Barendrecht on 25/07/88 hauling the 08.53 Amsterdam CS–Paris Nord service (EC 82 Etoile du Nord) formed of ex-TEE stainless steel stock. *David Brown*

▼Class 26 electric loco. No. 2607 (fitted with mono-motor bogies and variable-gear ratio) pulls out of Brussels Nord on 18/08/86 with a service bound for Luxembourg and formed of Type M4 coaching stock. These trains are now worked by 'break' units. *David Brown*

DIESEL RAILCARS

CLASS 43

The 1984 recast of services has put this class out of use, but some are still in use unrenumbered for departmental duties.

B (DMS).

Built: 1954–55.
Builder: Niv.
Wheel Arrangement: 2–B.
Engine: SEMG 6K103HS of 295 kW.
Transmission: Hydraulic. SEMG.
Accommodation: 95S 1L.
Total Weight: 40.5 tonnes.
Length over couplings: 23.80 m.
Max. Speed: 90 km/h.

Originally numbered 603.05–24.

4305 **D** FHS(D)
4306 **D** FKR(D)
4315 **D** FVS(D)
4319 **D** LL(D)
4324 **D** FCR(D)

CLASS 44

Used on branch lines, these units can work with class 734 trailers. The FKR units work Aalst–Burst whilst those at MUT are used on the Virton–Bastogne and Libramont–Dinant routes.

B (DMS).

Built: 1954.
Builder: Germain, rebuilt 1975–78 by CWFM.
Wheel Arrangement: B–2.
Engine: 2 x GM 6V71N of 118 kW each.
Transmission: Hydraulic. twin disc.
Max. Speed: 80 km/h.
Accommodation: 95S 1L.
Total Weight: 43 tonnes.
Length over couplings: 23.80 m.

Originally numbered 604.01–10.

4401 **D** FKR
4402 **D** FKR
4403 **D** FKR
4404 **D** FKR
4405 **D** FKR
4406 **D** FKR
4407 **D** MUT
4408 **D** MUT
4410 **D** MUT

CLASS 45

These units can work in multiple with or without class 734 trailers. Their use diminished with the timetable revisions of 1984 and they can now be found on the Virton–Bertrix–Libramont–Bastogne and Dinant–Libramont routes.

B (DMS).

Built: 1955.
Builder: Germain, rebuilt 1974–78 by CWFM.
Wheel Arrangement: 1A–A1.
Engine: 2 x GM 6V71N of 118 kW each.
Transmission: Hydraulic. Voith.
Max. Speed: 80 km/h.
Accommodation: 95S 1L.
Total Weight: 42 tonnes.
Length over couplings: 23.80 m.

Originally numbered 605.01–10.

4501 **D** MBX
4502 **D** MBX
4503 **D** MBX
4504 **D** MBX
4505 **D** MBX
4506 **D** MBX
4507 **D** MBX
4508 **D** MBX
4509 **D** MBX
4510 **D** MBX

CLASS 46

Another class with restricted use since the timetable changes. Those at GT are stored whilst those at GMN are used on the Montzen–Welkenraedt staff trains.

B (DMS).

Built: 1952.
Builder: Rag.
Accommodation: 74S 1L.

Wheel Arrangement: 1A–A1.
Engine: GM 6V71N of 118 kW.
Transmission: Hydraulic. Voith.
Total Weight: 24 tonnes.
Length over couplings: 16.22 m.
Max. Speed: 80 km/h.

Originally numbered 554.01–10.

4601 **D**	GT	4605 **D**	GT	4609 **D**	GMN	4610 **D**	GT
4603 **D**	GT	4608 **D**	GMN				

CLASS 49

The last survivor in traffic of a class that once totalled 50. The survivor is used for crew training trips from Antwerpen and railtours!

B (DMS).

Built: 1942.
Builder: Germain.
Wheel Arrangement: 1A–A1.
Engine: Brossel 8D120B of 113 kW.
Transmission: Mechanical.
Accommodation: 77S.
Total Weight: 22.7 tonnes.
Length over couplings: 15.985 m.
Max. Speed: 66 km/h.

Originally numbered 553.12.

4903 FNDM(D)

CLASS 734 TRAILER

Originally numbered 734.01–10, the original numbers are still carried and quoted in preference to the full UIC number. These trailers work as required with classes 44 and 45.

B (TS).

Built: 1955.
Builder: CWFM.
Accommodation: 80S 1L.
Total Weight: 17.45 tonnes.
Length over couplings: 16.11 m.
Max. Speed: 90 km/h.

UIC numbers are 50 88 27 29 580–589, with the end digit being one less than the end digit of the old number.

734.01	**D**	FKR	734.05	**D**	FKR	734.07	**D**	MBX
734.03	**D**	FKR	734.06	**D**	MBX	734.10	**D**	MBX
734.04	**D**	FKR						

▼Diesel railcar trailer No. 73.404 at Merelbeke on 26/03/89. M. Trower

DIESEL LOCOMOTIVES

CLASS 51 Co–Co

This class of mixed traffic locos is found all over the network except in the Liège and Luxembourg areas where class 55 is used. They are mostly used on freights but there are still some rush hour passenger trains from Brussels to De Panne and Ronse.

Built: 1961–63.
Builder-Mech. Parts: Cockerill.
Builder-Elec. Parts: ACEC/SEM.
Engine: Cockerill/Baldwin 10-608A of 1435 kW at 650 rpm.
Transmission: Electric. Six axle-hung traction motors.
Train Heating: Steam. Vapor OK4616.
Weight in Full Working Order: 117 (113.2, 5154–93) tonnes.
Maximum Tractive Effort: 272 kN. **Length over Buffers**: 20.160 m.
Driving Wheel Dia.: 1010 mm. **Max. Speed**: 120 km/h.

§–Fitted with carriage lighting jumpers for use with type M1 push & pull trains.
Originally numbered 200.002–093.

5101	**G**	FSR	5123	**G**	LNC	5145	**G**	FHS	5168	**G**	FNDM
5102	**G**	FSR	5124	**G**	FKR	5146	**G**	§ LNC	5170	**G**	FNDM
5103	**G**	§ FSR	5125	**G**	FKR	5147	**G**	LNC	5171	**G**	FNDM
5104	**G**	FSR	5126	**G**	LNC	5148	**G**	§ LNC	5172	**G**	FNDM
5105	**G**	FSR	5127	**G**	LNC	5149	**G**	LNC	5173	**G**	FNDM
5106	**G**	FSR	5128	**G**	FKR	5150	**G**	LNC	5174	**G**	FNDM
5107	**G**	FKR	5129	**G**	LNC	5151	**G**	FHS	5175	**G**	FNDM
5108	**G**	§ FKR	5130	**G**	LNC	5152	**G**	FHS	5177	**G**	FNDM
5109	**G**	FKR	5131	**G**	§ LNC	5153	**G**	FHS	5178	**G**	FNDM
5110	**G**	FKR	5132	**G**	FKR	5154	**G**	FNDM	5179	**G**	FHS
5111	**G**	FKR	5133	**G**	FHS	5155	**G**	FNDM	5180	**G**	FNDM
5112	**G**	FKR	5134	**G**	FKR	5156	**G**	FNDM	5181	**G**	FNDM
5113	**G**	FKR	5135		LNC	5157	**G**	FNDM	5182	**G**	FNDM
5114	**G**	FSR	5136	**G**	FHS	5158	**G**	FNDM	5183	**G**	FHS
5115	**G**	FKR	5137	**G**	FHS	5159	**G**	FNDM	5184	**G**	FHS
5116	**G**	FKR	5138	**G**	§ LNC	5160	**G**	FNDM	5185	**G**	FHS
5117	**G**	LNC	5139	**G**	LNC	5162	**G**	FNDM	5186	**G**	FHS
5118	**G**	FKR	5140	**G**	LNC	5163	**G**	FNDM	5187	**G**	FHS
5119	**G**	FKR	5141	**G**	LNC	5164	**G**	FNDM	5189	**G**	FHS
5120	**G**	LNC	5142	**G**	FHS	5166	**G**	FNDM	5192	**G**	FHS
5121	**G**	LNC	5143	**G**	FHS	5167	**G**	FNDM	5193	**G**	FHS
5122	**G**	FSR	5144	**G**	FHS						

CLASS 52 Co–Co

Classes 52, 53 and 54 are part of a large European family of locomotives. The design originated as Nohab/GM and similar locomotives are found in Denmark (classes MV, MX and MY), Hungary (class M61) and Luxembourg (class 1600). Class 52 can be found on passenger work between Namur and Dinant and on freight work between Namur and Stockem via Virton. After many complaints from crews new cabs have been fitted which substantially alter the appearance of the locos. Since the restructuring of passenger services several locos have lost their boilers and been have reclassified to class 53.

Built: 1955.
Builder-Mech. Parts: AFB.
Builder-Elec. Parts: GM.
Engine: GM 16-567C of 1265 kW at 835 rpm.
Transmission: Electric. 6 axle-hung traction motors.
Train Heating: Steam. Vapor OK4616. **Weight in Full Working Order**: 108 tonnes.
Maximum Tractive Effort: 245 kN. **Length over Buffers**: 18.850 m.
Driving Wheel Dia.: 1010 mm. **Max. Speed**: 120 km/h.

Rheostatic braking.
*–Fitted with new cabs.

▲Class 28 No. 2802 stands at Brussels Midi whilst on station pilot duty on 03/08/88. *S.C. Falcus*

▼Class 51 No. 5103 at Brussels Midi with the 16.33 to De Panne on 24/04/89 formed of 7 M2 coaches. *Brian Garvin*

The original 5201–13 were originally numbered 202.001–013.
5214–17 were renumbered from 5307/02/17/18.

5201 **G** * FEO	5209 **G** MUT	5213 **G** * FEO	5216 **G** * FEO
5202 **G** * FEO	5211 **G** * FEO	5214 **G** * FEO	5217 **G** * MUT
5205 **G** * FEO	5212 **G** * FEO	5215 **G** * FEO	

CLASS 53 Co–Co

These are similar to class 52 but have no train heating. Similar freight use to class 52. Several class 52s have recently lost their boilers and been renumbered to class 53.

Built: 1955.
Builder-Mech. Parts: AFB.
Builder-Elec. Parts: GM.
Engine: GM 16-567C of 1265 kW at 835 rpm.
Transmission: Electric. 6 axle-hung traction motors.
Train Heating: None.
Maximum Tractive Effort: 245 kN.
Driving Wheel Dia.: 1010 mm.
Weight in Full Working Order: 106.6 tonnes.
Length over Buffers: 18.850 m.
Max. Speed: 120 km/h.

Rheostatic braking.
* Fitted with new cabs.
The original 5301–19 were originally numbered 203.001–019.
5302/07/17/18/20 have been renumbered from 5203/06/07/08/10.

5301 **G** * MUT	5306 **G** * MUT	5311 **G** * MUT	5316 **G** * MUT
5302 **G** * FEO	5307 FEO	5312 **G** * MUT	5317 **G** * FEO
5303 **G** * MUT	5308 **G** * MUT	5313 **G** * MUT	5318 FEO
5304 **G** * MUT	5309 **G** * MUT	5314 **G** * MUT	5319 MUT
5305 **G** * MUT	5310 **G** * MUT	5315 **G** * MUT	5320 FEO

CLASS 54 Co–Co

Similar to class 52, but no rheostatic braking and an additional headlight. Similar use to class 52. None have yet been fitted with new cabs and all are expected to be withdrawn by 1988. 5403 and 5404 carry original liveries and numbers. 5404 is likely to be preserved.

Built: 1955–57.
Builder-Mech. Parts: AFB.
Builder-Elec. Parts: GM.
Engine: GM 16-567C of 1265 kW at 835 rpm.
Transmission: Electric. 6 axle-hung traction motors.
Train Heating: Steam. Vapor OK4616.
Maximum Tractive Effort: 245 kN.
Driving Wheel Dia.: 1010 mm.
Weight in Full Working Order: 108 tonnes.
Length over Buffers: 18.850 m.
Max. Speed: 120 km/h.
Originally numbered 204.001–007.

5401 FEO	5403 FEO	5404 FEO	5407 FEO

CLASS 55 Co–Co

This class is based in the Liège area for freight and passenger work in the Ardennes. Those with electric heating are used on certain trains on the Liège–Luxembourg line and are painted in blue & yellow livery instead of the standard yellow & green.

Built: 1961–62.
Builder-Mech. Parts: BN.
Builder-Elec. Parts: ACEC/SEMG.
Engine: GM 16-567C of 1435 kW at 835 rpm.
Transmission: Electric. Six axle-hung traction motors.
Train Heating: Steam. Vapor 4616.
Maximum Tractive Effort: 272 kN.
Driving Wheel Dia.: 1010 mm.
Weight in Full Working Order: 110 tonnes.
Length over Buffers: 19.550 m.
Max. Speed: 120 km/h.

Rheostatic braking.
e–Electric heating fitted using an ACEC 300 kW alternator.
§–Fitted with carriage lighting jumpers for use on M1 type push & pull trains.
Originally numbered 205.001–042.

▲Rebuilt Class 52 diesel electric No. 5212 approaches Namur station on 12/09/87. *Colin Boocock*

▼Class 59s Nos. 5936 and 5946 pass Leuven with a freight on 29/07/88. *Alex Dasi-Sutton*

5501 . MUT	5511 . NK	5523 **Y** e NK	5533 **G** § NK
5502 . MUT	5512 . § NK	5524 . § NK	5534 **G** NK
5503 . MUT	5513 NK	5525 **G** § NK	5535 NK
5504 MUT	5514 NK	5526 **G** NK	5536 **G** NK
5505 **Y** e NK	5515 **Y** e NK	5527 **G** NK	5537 § NK
5506 NK	5517 **G** § NK	5528 **G** NK	5538 § NK
5507 NK	5518 **G** NK	5529 **Y** e NK	5539 **G** NK
5508 NK	5519 . § NK	5530 **G** NK	5540 **Y** e NK
5509 . NK	5520 **G** NK	5531 § NK	5541 . § NK
5510 **Y** e NK	5521 **G** § NK	5532 **G** NK	5542 **Y** e NK

CLASS 59 — Bo–Bo

These locos are used on freights from Antwerpen and due for early withdrawal. Some may be retained for use on LGV construction trains.
Built: 1955.
Builder-Mech. Parts: Cockerill/ BN/Niv.
Builder-Elec. Parts: ACEC.
Engine: Cockerill/Baldwin 608A of 1280 kW at 625 rpm.
Transmission: Electric. Four axle-hung traction motors.
Train Heating: Steam. Vapor 4616.
Maximum Tractive Effort: 196 kN.
Driving Wheel Dia.: 1110 mm.
Weight in Full Working Order: 87.2 tonnes.
Length over Buffers: 16.180 m.
Max. Speed: 120 km/h.

Originally numbered 201.005–046.
5910 carries old livery and number 201.010.

5905 FNDM	5910 FNDM	5936 **G** FNDM	5946 FNDM
5909 **G** FNDM	5926 **G** FNDM		

CLASS 60 — Bo–Bo

These few locos are in departmental use by the permanent way department.

Built: 1963–66.
Builder-Mech. Parts: Cockerill.
Builder-Elec. Parts: ACEC/SEMG.
Engine: Cockerill TH8.95SA of 1030 kW at 1000 rpm.
Transmission: Electric. Four axle-hung traction motors.
Train Heating: Steam. Vapor OK46516
Maximum Tractive Effort: 196 kN.
Driving Wheel Dia.: 1010 mm.
Weight in Full Working Order: 78 tonnes.
Length over Buffers: 17.150 m.
Max. Speed: 120 km/h.

Originally numbered 210.016–089.

6016 . FKR(D)	6073 . FKR(D)	6086 . FKR(D)	6089 **G** FGH(D)
6042 **G** FGH(D)			

CLASS 62 — Bo–Bo

These locos are used on mixed traffic duties. They are found on passenger work around Gent, Mons, Hasselt and Antwerpen. 6391–93 were built first as prototypes with flexicoil suspension, but the production series have bogies designed by BN. A few locos are already scrapped following accidents.

Built: 1961–66.
Builder-Mech. Parts: BN.
Builder-Elec. Parts: ACEC.
Engine: GM 12-567C of 1050 kW at 835 rpm.
Transmission: Electric. Four axle-hung traction motors.
Train Heating: Steam. Vapor OK4616.
Maximum Tractive Effort: 190 kN.
Driving Wheel Dia.: 1010 mm.
Weight in Full Working Order: 80 (78*) tonnes.
Length over Buffers: 16.790 m.
Max. Speed: 120 km/h.

§–Fitted with carriage lighting jumpers for M1 push & pull trains.
e–Fitted with electric heat instead of steam.
m–Museum loco livery.
Originally numbered 212.101–133, 212.001–003.

▲A pair of Class 62s Nos. 6319 and 6240 head out of Leuven Yard with a freight train on 29/07/88. *Alex Dasi-Sutton*

▼Class 70 shunter No. 7006 stands at Antwerpen Dam depot on 27/03/89. *M. Trower*

6201 **G** FKR
6202 **G** FKR
6203 **G** FKR
6204 **G** FKR
6205 **G** FKR
6206 **G** FKR
6207 **G** FKR
6210 **G** FKR
6211 **G** FKR
6212 **G** FKR
6213 **G** FKR
6214 **G** FKR
6215 **Y** e LK
6216 **G** FKR
6217 **G** § LNC
6218 **G** § LNC
6219 **G** FKR
6220 **G** FKR
6221 **G** FKR
6222 **G** § FKR
6223 **G** FKR
6224 **G** § LNC
6225 **G** FKR
6227 **G** FKR
6228 **G** FKR
6229 **G** FKR
6230 **G** FKR
6231 **G** § LNC
6233 **G** FKR
6234 **G** § LNC
6235 **G** FKR
6236 **G** FKR
6237 **G** FKR
6238 **G** FHS
6240 **G** FHS
6241 **G** FGH
6242 **G** FGH
6243 **G** § LNC
6244 FGH
6245 **G** FHS
6246 **G** FHS
6247 **G** LK
6248 **G** § LNC
6249 **G** § LNC
6250 FGH
6251 **G** FHS
6252 **G** FHS
6253 **G** FHS
6254 FGH
6255 **G** FHS
6256 **G** § FKR
6257 **G** FGH
6258 **G** FGH
6260 **G** FHS
6261 **G** FHS
6262 FGH
6263 FGH
6264 **G** FHS
6265 **G** § FKR
6266 **G** FGH
6267 **G** FHS
6268 **G** FHS
6269 FGH
6270 **G** FGH
6271 **G** § LNC
6272 FGH
6273 FGH
6274 **G** § NK
6275 **G** § FGH
6276 **G** FGH
6277 **G** FGH
6278 **G** FGH
6279 **G** FHS
6281 **G** FHS
6282 **G** FGH
6283 **G** FHS
6284 **G** FGH
6285 **G** FGH
6286 **G** FGH
6287 **G** FGH
6288 **G** FGH
6289 § NK
6290 **G** FGH
6291 **G** FHS
6292 **G** FKR
6293 **G** FKR
6294 **G** FHS
6295 **G** FHS
6296 **G** FHS
6297 **G** FHS
6298 **G** FHS
6299 **G** FHS
6300 **G** LK
6301 **G** LK
6302 LK
6303 **G** LK
6304 **G** LK
6305 **G** LK
6306 **G** LK
6307 **G** LK
6309 **G** FKR
6311 **G** FKR
6312 **G** FKR
6313 **G** FKR
6314 **G** FGH
6315 **G** FKR
6316 **G** FHS
6317 **G** FHS
6318 **G** FHS
6319 **G** FHS
6320 **G** FGH
6321 **G** FHS
6322 **G** FKR
6323 § NK
6324 **G** § NK
6325 **G** NK
6326 **G** FHS
6327 **G** FHS
6328 **G** NK
6329 § NK
6330 **G** FHS
6331 FHS
6333 **G** FKR
6391 **G** * FKR
6392 **G** * FKR
6393 **G** * FKR

CLASS 70 Bo–Bo

This class is used on heavy shunting duties in Antwerpen Docks.

Built: 1954.
Builder-Mech. Parts: BM.
Builder-Elec. Parts: ACEC.
Engine: ABC 8DUS of 515 kW at 650 rpm (ABC 6DXC of 550 kW at 750 rpm*).
Transmission: Electric. Four axle-hung traction motors.
Train Heating: None.
Maximum Tractive Effort: 196 kN.
Driving Wheel Dia.: 1070 mm.
Weight in Full Working Order: 84.6 tonnes.
Length over Buffers: 12.150 m.
Max. Speed: 50 km/h.

Originally numbered 222.001–003.

7001 **G** * FNDM
7002 **G** FNDM
7003 **G** FNDM
7004 **G** FNDM
7005 **G** FNDM
7006 **G** FNDM

CLASS 71 B–B

Rebuilt from 6601–03 in 1980 at FAZ. Used for heavy shunting duties in Antwerpen Docks.

Built: 1962.
Builder: ABR.
Engine: ABC 6DXC 100-750A of 662 kW at 750 rpm.
Transmission: Hydraulic. Voith L217.
Train Heating: None.
Maximum Tractive Effort: 197 kN.
Driving Wheel Dia.: 1010 mm.
Weight in Full Working Order: 74 tonnes.
Length over Buffers: 13.440 m.
Max. Speed: 80 km/h.

a–Fitted with automatic coupler

Originally numbered 271.001/002/006.

7101 **G** FNDM
7102 **G** FNDM
7103 **G** a FNDM

CLASS 73 — C

General purpose shunters.

Built: 1965–66 (1973–74*, 1976–77†).
Builder: BN (ABR§).
Engine: Cockerill 6TH695SA (6T240CO*†) of 550 kW at 950 rpm.
Transmission: Hydraulic. Voith L217u.
Train Heating: None.
Maximum Tractive Effort: 211 kN.
Driving Wheel Dia.: 1262 mm.
Weight in Full Working Order: 56 tonnes.
Length over Buffers: 11.170 (11.404*†) m.
Max. Speed: 60 km/h.

a Auto-couplers for shunting.
† Multiple working fitted for use in hump yards.
7301–35 were originally numbered 273.001–035.

No.		Depot	No.		Depot	No.		Depot	No.		Depot
7301		LNC	7325		LNC	7349	*	NK	7373	*a	FHS
7302		LNC	7326	§	LNC	7350	*	FKR	7374	*	FHS
7303		LNC	7327	§	LNC	7351	*	FKR	7375	*a	FHS
7304		LNC	7328	§	LNC	7352	*	FKR	7376	†	FKR
7305		LNC	7329	§	LNC	7353	*	FKR	7377	†	FKR
7306		LNC	7330	§	LNC	7354	*	FKR	7378	†	FKR
7307		LNC	7331	§	LNC	7355	*	FKR	7379	†	FKR
7308		LNC	7332	§	LNC	7356	*	FKR	7380	†	FKR
7309		LNC	7333	§	LNC	7357	*a	FHS	7381	†	NK
7310		LNC	7334	§	LNC	7358	*	FKR	7382	†	NK
7311		LNC	7335	§	LNC	7359	*	FKR	7383	†	NK
7312		LNC	7336	*	FHS	7360	*	FKR	7384	†	NK
7313		LNC	7337	*	FKR	7361	*	FKR	7385	†	NK
7314		LNC	7338	*	FEO	7362	*	FKR	7386	†	NK
7315		LNC	7339	*	FEO	7363	*	FKR	7387	†	FHS
7316		LNC	7340	*	FEO	7364	*	FKR	7388	†	NK
7317		LNC	7341	*	FEO	7365	*a	FHS	7389	†	NK
7318		LNC	7342	*	FEO	7366	*	FHS	7390	†	NK
7319		LNC	7343	*	NK	7367	*	FKR	7391	†	NK
7320		LNC	7344	*	NK	7368	*	FKR	7392	†	FHS
7321		LNC	7345	*	NK	7369	*	FKR	7393	†	FHS
7322		LNC	7346	*	NK	7370	*	FKR	7394	†	FHS
7323		LNC	7347	*	NK	7371	*a	FHS	7395	†	FHS
7324		LNC	7348	*	NK	7372	*	FKR			

CLASS 74 — C

Shunters used in yards around Antwerpen.

Built: 1977.
Builder: BN.
Engine: ABC 6DXS of 550 kW at 750 rpm.
Transmission: Hydraulic. Voith L217u.
Train Heating: None.
Maximum Tractive Effort: 196 kN.
Driving Wheel Dia.: 1262 mm.
Weight in Full Working Order: 59 tonnes.
Length over Buffers: 11.404 m.
Max. Speed: 60 km/h.

Multiple working fitted.

No.	Depot	No.	Depot	No.	Depot	No.	Depot
7401	FNDM	7404	FNDM	7407	FNDM	7409	FNDM
7402	FNDM	7405	FNDM	7408	FNDM	7410	FNDM
7403	FNDM	7406	FNDM				

CLASS 75 — B–B

Originally main-line locos of class 65, these were downgraded to shunters in 1982–83 when their steam boilers and water tanks were removed. They are now used for heavy trip working around Antwerpen Docks.

Built: 1965.
Builder: BN.
Engine: GM 12-567D-1 (two-stroke) of 1075 kW at 835 rpm.
Transmission: Hydraulic. Voith L216 rsb.

Train Heating: None.
Maximum Tractive Effort: 191 kN.
Driving Wheel Dia.: 1118 mm.
Originally numbered 203.001–006.
Weight in Full Working Order: 79 tonnes.
Length over Buffers: 16.790 m.
Max. Speed: 120 km/h.

7501 (6501)	**G**	FNDM	7503 (6503)	**G**	FNDM	7505 (6505)	**G** FNDM
7502 (6502)	**G**	FNDM	7504 (6504)	**G**	FNDM	7506 (6506)	**G** FNDM

CLASS 80 C

General purpose shunter used around Brussels.

Built: 1960–1963.
Builder: BN (ABR*).
Engine: Maybach GTO6A of 480 kW at 1400 rpm.
Transmission: Hydraulic. Voith L37z Ub.
Train Heating: None.
Maximum Tractive Effort: 173 kN.
Driving Wheel Dia.: 1252 mm.
Originally numbered 260.001–069.
Weight in Full Working Order: 52 tonnes.
Length over Buffers: 10.360 m.
Max. Speed: 60 km/h.

8001	FSR	8025	* FSR	8046	FSR	8061	FSR
8002	FSR	8027	* FSR	8047	FSR	8062	FSR
8006	FSR	8031	* FSR	8049	FSR	8063	FSR
8007	FSR	8032	* FSR	8050	FSR	8064	FSR
8008	FSR	8033	* FSR	8051	FSR	8065	FSR
8009	FSR	8034	* FSR	8052	FSR	8066	FSR
8011	FSR	8035	* FSR	8053	FKR	8067	FSR
8012	FSR	8037	* FSR	8055	FSR	8068	FSR
8018	FSR	8040	* FSR	8058	FSR	8069	FSR
8020	FSR	8045	FSR	8059	FSR		

CLASS 82 C

General purpose shunters.

Built: 1965–73.
Builder: BN.
Engine: ABC 6DXS of 480 kW at 750 rpm.
Transmission: Hydraulic. Voith L217u.
Train Heating: None.
Weight in Full Working Order: 57 (59 8256–74, 56 s) tonnes.
Maximum Tractive Effort: 191 kN.
Driving Wheel Dia.: 1262 mm.
Length over Buffers: 11.170 (11.320 s) m.
Max. Speed: 60 km/h.

m–Fitted for multiple working around Antwerpen.
s–Rebuilt in 1977 to a slave unit by the removal of the cab.
8201–55 were originally numbered 262.001–055.

8201	FSD	8220	FVY	8239	NK	8258	m FNDM
8202	FSD	8221	FGH	8240	NK	8259	m FNDM
8203	FVY	8222	FGH	8241	FGH	8260	m FNDM
8204	FSD	8223	FGH	8242	FSD	8261	m FNDM
8205	NK	8224	NK	8243	FSD	8262	m FNDM
8206	FSD	8225	NK	8244	FSD	8263	m FNDM
8207	FSD	8226	NK	8245	NK	8264	m FNDM
8208	FSD	8227	FGH	8246	FNDM	8265	m FNDM
8209	FSD	8228	FGH	8247	FNDM	8266	m FNDM
8210	FSD	8229	FGH	8248	NK	8267	m FNDM
8211	FSD	8230	FGH	8249	FSD	8268	m FNDM
8212	FGH	8231	FGH	8250	FSD	8269	m FNDM
8213	NK	8232	FNDM	8251	FGH	8270	m FNDM
8214	FGH	8233	FSD	8252	NK	8271	m FNDM
8215	NK	8234	NK	8253	NK	8272	m FNDM
8216	NK	8235	NK	8254	FGH	8273	m FNDM
8217	FGH	8236	FNDM	8255	NK	8274	m FNDM
8218	FGH	8237	NK	8256	m FNDM	8275	s FNDM
8219	FGH	8238	NK	8257	m FNDM		

CLASS 83 C

All based at Monceau for shunting the yards around the Charleroi industrial area.

Built: 1956.
Builder: Cockerill.
Engine: Cockerill 695 SA of 405 kW at 450 rpm.
Transmission: Hydraulic. Voith L37u.
Train Heating: None.
Maximum Tractive Effort: 157 kN.
Driving Wheel Dia.: 1262 mm.
Weight in Full Working Order: 57 tonnes.
Length over Buffers: 10.750 m.
Max. Speed: 50 km/h.

Originally numbered 253.001–025.

8301	LNC	8310	LNC	8315	LNC	8320	LNC
8302	LNC	8311	LNC	8316	LNC	8322	LNC
8303	LNC	8312	LNC	8317	LNC	8323	LNC
8304	LNC	8313	LNC	8319	LNC	8325	LNC
8305	LNC	8314	LNC				

CLASS 84 C

General purpose shunters. 8461–70 were renumbered from 8526–8535.

Built: 1955–56*, 1962–63, 1959 reb 1968–79§.
Builders: ABR (BM*§).
Engine: ABC 6DUS (6DXS§) of 405 kW at 680 rpm.
Transmission: Voith L37u.
Train Heating: None.
Weight in Full Working Order: 55.8 (54.2*, 57.3§) tonnes.
Maximum Tractive Effort: 157 kN.
Length over Buffers: 10.650 (10.150*§) m.
Driving Wheel Dia.: 1262 mm.
Max. Speed: 50 km/h.
Originally numbered 250.004–025, 250.101–135, 252.026–035.

8404	*GMN	8424	*GMN	8441	FSD	8457	FNDM
8405	*GMN	8426	FSD	8442	FNDM	8458	FNDM
8406	*GMN	8427	MUT	8443	LK	8459	FNDM
8407	*GMN	8428	GT	8444	LK	8460	FNDM
8408	*GMN	8429	FSD	8447	GT	8461	§FNDM
8410	*GT	8430	FSD	8448	GT	8462	§FNDM
8412	*GMN	8431	FSD	8449	GT	8463	§FNDM
8414	*GMN	8432	FSD	8450	GT	8464	§FNDM
8415	*MUT	8433	GMN	8451	FNDM	8465	§FNDM
8417	*GT	8434	FSD	8452	FNDM	8466	§FNDM
8418	*MUT	8435	LK	8453	FNDM	8467	§FNDM
8419	*MUT	8437	FSD	8454	FNDM	8468	§FNDM
8421	*MUT	8439	LK	8455	FNDM	8469	§FNDM
8422	*MUT	8440	LK	8456	FNDM	8470	§FNDM

CLASS 85 C

Used for general purpose shunting around Antwerpen.

Built: 1956–57.
Builder: FUF.
Engine: ABC 6DXS of 405 kW at 680 rpm.
Transmission: Voith L37u.
Train Heating: None.
Maximum Tractive Effort: 157 kN.
Driving Wheel Dia.: 1262 mm.
Weight in Full Working Order: 57.3 tonnes.
Length over Buffers: 10.000 m.
Max. Speed: 50 km/h.

Originally numbered 252.001–025.

8501	FNDM	8508	FNDM	8514	FNDM	8520	FNDM
8502	FNDM	8509	FNDM	8515	FNDM	8521	FNDM
8503	FNDM	8510	FNDM	8516	FNDM	8522	FNDM
8504	FNDM	8511	FNDM	8517	FNDM	8523	FNDM
8505	FNDM	8512	FNDM	8518	FNDM	8524	FNDM
8506	FNDM	8513	FNDM	8519	FNDM	8525	FNDM
8507	FNDM						

CLASS 91 B

This class was originally numbered 9001–60 and was rebuilt in the late 1970s with more powerful engines and in some cases lengthened frames ready for automatic couplings! Many have been made spare by changes in freight workings and have been transferred to departmental use. Those in traffic stock are used for shunting at small stations and yards. Some are used as loco depot pilots.

Built: 1961–64.
Builders: Cockerill (9101–10), ABR (9111–9135), Niv (9136–60).
Engine: GM 12V71N (two-stroke) of 245 kW at 1800 rpm.
Transmission: Hydraulic. Twin Disc 11500 HS390.
Train Heating: None
Maximum Tractive Effort: 90 kN.
Driving Wheel Dia.: 920 mm.
Weight in Full Working Order: 35 tonnes.
Length over Buffers: 8.055 m.
Max. Speed: 40 km/h.

Originally numbered 230.001–010/101–150.

9101	FGSP(D)	9116	FEO	9131	FGH	9146	FEO
9102	FGSP(D)	9117	FGH	9132	FKR	9147	FEO
9103	FGSP(D)	9118	FGH	9133	FGH	9148	MKM
9104	LK	9119	FSR(D)	9134	MKM	9149	FEO
9105	LK	9120	FGH	9135	MKM	9150	FWO(D)
9106	LK	9121	FSR(D)	9136	FKR	9151	FEO
9107	LK	9122	FMS(D)	9137	MKM	9152	FKR
9108	LK	9123	FKR	9138	FGSP(D)	9153	MKM
9109	LK	9124	FMS(D)	9139	FNDM	9154	FNR(D)
9110	LK	9125	MKM	9140	LBC(D)	9155	FEO
9111	FGH	9126	FGH	9141	FCR(D)	9156	FEO
9112	FGH	9127	MKM	9142	FNR(D)	9157	FEO
9113	FGH	9128	FMS(D)	9143	FEO	9158	FEO
9114	FGH	9129	FNDM	9144	FGH	9159	MKM
9115	FGH	9130	FGH	9145	MGR(D)	9160	FKR

CLASS 92 C

This class is now wholly in departmental use and used by the permanent way department on ballast trains etc. The locos are often stabled in engineers depots.

Built: 1960.
Builder: BN.
Engine: SEM 6K113HS of 255 kW at 1300 rpm.
Transmission: Hydraulic. Voith L37u.
Train Heating: None.
Maximum Tractive Effort: 147 kN.
Driving Wheel Dia.: 1262 mm.
Weight in Full Working Order: 50.5 tonnes.
Length over Buffers: 10.40 m.
Max. Speed: 45 km/h.
Originally numbered 232.001–025.

9201	FKR(D)	9208	FFO(D)	9214	LNC(D)	9220	FFO(D)
9202	FFO(D)	9209	FKR(D)	9215	MGR(D)	9221	FFO(D)
9203	FGH(D)	9210	MGR(D)	9216	LNC(D)	9222	FHS(D)
9204	FNDM(D)	9211	MGR(D)	9217	FEO(D)	9223	FGH(D)
9205	FHS(D)	9212	FNDM(D)	9218	FKR(D)	9224	FEO(D)
9206	MGR(D)	9213	FEO(D)	9219	FNDM(D)	9225	FKR(D)
9207	FEO(D)						

SELF-PROPELLED DEPARTMENTAL STOCK

Mention has already been made of classes 91 and 92 being used for departmental traffic. The other self-propelled departmental stock consists mainly of overhead line inspection units. Full details of these are not available. They are found in separate sheds belonging to the traction supply department.

ES100 SERIES

Built new 1947–49.

ES 101	FSR	ES 103	GNS	ES 105	NK	ES 106	FKR
ES 102	FSR	ES 104	FSR				

8246

9217

8033

9116

ES200 SERIES

Built new 1971–72.

ES 201	FCR	ES 204	**R**	FKR	ES 207	**R**	FMS	ES 210		FR
ES 202	FSR	ES 205		LL	ES 208		FLV	ES 211		FNR
ES 203	LJ	ES 206		GNS	ES 209		NK	ES 212		FSR

ES400 SERIES

Conversions from class 43 DMUs.

ES 401	(4307)	FM	ES 402	(4325)	FMS

ES4600/4900 SERIES

Converted from class 46/49 DMUs and not renumbered.

4612	FVS	4905	LJ	4907	FNR	4911	FLV
4901	FVS						

SNCB WITHDRAWN LOCOMOTIVES

070	**FM**	4607	FM	6001	FAZ	6054	Schlessin	8014	(FKR)
087	FM	4613	GT	6003	FAZ	6055	FEO	8015	FSR
089	FM	4615	GT	6004	FAZ	6056	FHS	8016	FSR
513	FM	4617	GT	6005	FAZ	6058	FVY	8017	FSR
533	FM	4619	GT	6006	FAZ	6059	FHS	8021	(FKR)
901	FM	5204	FAZ	6010	FGH	6060	(FKR)	8022	Italy
903	FM	5901	FNDM	6011	FHS	6061	FAZ	8023	(FKR)
904	FM	5902	FAZ	6012	Italy	6062	FEO	8024	FSR
2906	FEO	5903	(FNDM)	6013	FAZ	6063	FEO	8026	FSR
2909	FEO	5906	(FKR)	6014	FEO	6064	FDR	8028	FSR
2910	FEO	5907	FNDM	6018	FAZ	6065	FVY	8029	(FKR)
2920	FEO	5911	(FKR)	6019	GT	6066	FVY	8036	FSR
4002	FM	5912	FNDM	6020	FAZ	6067	FGH	8038	(FKR)
4003	FM	5913	(FNDM)	6021	FAZ	6068	GT	8039	FSR
4004	FM	5914	FNDM	6022	FEO	6069	FGH	8041	FSR
4005	FM	5916	FNDM	6023	FEO	6070	FGH	8042	(FKR)
4007	FM	5917	FNDM	6024	FEO	6071	FEO	8043	(FSR)
4301	GMN	5918	FKR	6025	FAZ	6072	FAZ	8044	(FSR)
4302	(LK)	5919	FNDM	6026	FEO	6074	FHS	8048	Italy
4304	GMN	5921	FNDM	6027	(FKR)	6075	FAZ	8054	(FSR)
4305	FHS	5922	FKR	6028	FGH	6076	FEO	8056	FSR
4306	FKR	5923	FNDM	6029	FVY	6077	FAZ	8057	FSR
4308	LK	5924	FNDM	6030	FHS	6078	FAZ	8060	Mariembourg Ind.
4309	(LK)	5925	FNDM	6032	FGH	6079	FKR	8306	FAZ
4310	GMN	5928	FKR	6033	FVY	6081	FEO	8307	LNC
4312	MBX	5929	FKR	6034	(FGH)	6082	FKR	8308	LNC
4313	LK	5930	FNDM	6035	FVY	6083	FAZ	8309	LNC
4314	MUT	5931	FNDM	6036	FGH	6085	FEO	8321	LNC
4316	MUT	5933	Italy	6037	GT	6088	Italy	8324	LNC
4318	GMN	5935	FNDM	6038	FGH	6090	FEO	8401	(MUT)
4320	LK	5937	FNDM	6039	Italy	6101	FDR	8402	FAZ
4321	MUT	5938	FNDM	6040	FGH	6102	FDR	8403	MUT
4322	MUT	5939	FKR	6041	FVY	6104	FAZ	8409	(GMN)
4326	(LK)	5940	FKR	6043	FEO	6105	FDR	8410	GT
4327	MUT	5941	FNDM	6044	FEO	6106	FDR	8411	Italy
4328	(LK)	5943	FNDM	6045	Italy	6109	FAZ	8413	(MKM)
4330	MUT	5944	FKR	6046	FGH	6110	FDR	8416	GT
4331	FM	5945	FNDM	6047	(FGH)	6111	FAZ	8420	(MUT)
4332	MBX	5947	FKR	6048	FGH	6114	FDR	8423	(GMN)
4333	MUT	5949	FNDM	6049	FEO	6115	FDR	8425	(MUT)
4334	MUT	5950	FNDM	6050	(FGH)	8004	FSR	8436	Italy
4336	MUT	5951	FKR	6051	FEO	8005	Italy	8438	Italy
4409	FM	5952	FNDM	6052	FGH	8010	FSR	8445	Italy
4602	(ATH)	5953	FKR	6053	FEO	8013	Italy	8446	Italy
4606	GT	5954	FNDM						

BELGIAN PRESERVED LOCOS & RAILCARS

STATUS CODES

A Active (location could vary).
P Plinthed
7 Retained for special excursions.
M Museum or Museum line loco.
R Under restoration (perhaps at another place).
S Stored or for spares.

STEAM LOCOMOTIVES:

Number	*Type*	*Built*	*Status*	*Location*
1.002	4–6–2	1921	MR	FLV
7.039	4–6–0	1921	MS	FLV
10.018	4–6–2	1911	MS	FLV
12.004	4–4–2	1939	MA	FLV
16.042	4–4–2T	1909	MS	FLV
18.051	4–4–0	1905	MS	FLV
29.013	2–8–0	1945	MA	FLV
41.195	0–6–0	1910	P	FC
44.021	0–6–0	1906	MS	FLV
44.225	0–6–0	1907	MS	FLV
53.320	0–8–0T	1906	MS	FLV
64.045	4–6–0	1918	MS	FLV
2	2–2–2ST	1842	M	Brussels Nord
615	0–8–0T	1859	MS	FLV
1151	0–6–0PT	1879	MS	FLV
3364	0–4–0VB	1877	MP	Steamtown USA

DIESEL & ELECTRIC LOCOMOTIVES:

Number	*Wheels*	*Built*	*Status*	*Location*
2912	Bo–Bo e	1949	MS	FLV
2913	Bo–Bo e	1949	P	CWFM
5910	Bo–Bo de	1955	MA	FNDM. Restored as 210.010
5927	Bo–Bo de	1955	MA	SDP Baasrode
5930	Bo–Bo de	1955	MA	FNDM. Restored as 201.001
6406	BB dh	1962	MS	FLV
7103	D dh	1957	MS	FLV
7209	D dh	1956	MS	FNDM

DIESEL & ELECTRIC RAILCARS:

Number	*Wheels*	*Built*	*Status*	*Location*
4001	3-car DHMU	1957	MA	CFV3V
4006	3-car DHMU	1957	MS	LK
4602	1A–A1 DHMU	1952	MA	SDP Baasrode
4604	1A–A1 DHMU	1952	P	ULM Isières
4611	1A–A1 DHMU	1952	MA	CFV3V Mariembourg
4614	1A–A1 DHMU	1952	MA	FSR. Restored as 554.18
4616	1A–A1 DHMU	1952	MA	CFV3V Mariembourg
4618	1A–A1 DHMU	1952	MA	MSTB Vilvoorde
4620	1A–A1 DHMU	1952	MA	MSTB Baasrode
4906	1A–A1 DMMU	1942	MA	CFV3V
551.48	B DMMU	1939	MS	FTY
608.05	1A–A1 DMMU	1939	MS	CWFM
654.02	1/3 DEMU	1936	MS	FLV
002	2-car EMU	1939	MS	FLV
7.312/ 7.724	4-car EMU	1935	MA	FSR

Photos on page 38: VARIOUS SNCB SHUNTERS. Top left: 8033 at Brussels Midi. Top right: 8246 at Antwerpen Centraal, 02/06/89. (both P. Fox). Bottom left: 9116 at Namur. Bottom right: 9217 at Namur, 12/09/87. (both C. Boocock).

HAULED COACHING STOCK

Numbering. All Belgian stock has the official UIC number painted on the side of the coach in the centre, and its 'old' number, duly crossed out, either on the bottom left-hand corner or on the end of the vehicle. All stock has an 'old' number, including new stock being delivered! Old numbers are used in this section, with a note on the UIC series in the class details.

Types. There are three basic categories of hauled stock in Belgium. 'I' stock is for international use and has 2+1 seating in open firsts, 2+2 seating in open seconds and 6 seats to both first and second class compartments. 'K' stock is old slam-door stock and has hopper windows (K1) or wind-down windows (K2/3). One line on which it is to be found is Namur–Dinant. 'M' stock is sliding door stock. M1/2/3 is old design with reduced height. M4 stock is more modern, having tinted windows and pressure ventilation. M5 stock is double-deck push-pull stock used around Brussels in the peak periods. Seating in 'K' and 'M' stock is 2+2 in first class and 2+3 in second class. It must be stated that comfort is extremely poor in all Belgian non-international coaching stock.

Liveries. Livery on K and M1/2/3 stock is plain green, whereas M4/5 stock have the new Bordeaux red livery. However, M2/3 stock are now being repainted in red livery, but no details are available as to which coaches have been so-painted. Similarly, new I6 and I10 stock is in orange livery with a white band. Older international stock has been repainted in this colour, but it is possible that certain vehicles may still retain green livery. Livery codes are not shown in this section.

INTERNATIONAL STOCK

VARIOUS SPECIAL VEHICLES

No details are available of these various special vehicles, although it is known that 31 and 32 are medical vehicles. This list shows both the old and computer numbers.

11	60 88 99 70 011-6	52	60 88 99 69 052-3	63	60 88 99 69 103-5
13	60 88 99 70 013-2	53	60 88 99 69 053-1	64	60 88 99 69 104-3
31	60 88 80 69 031-8	54	60 88 99 69 054-9	65	60 88 99 69 105-0
32	60 88 99 69 032-6	55	60 88 99 69 055-6	71	60 88 99 09 001-3
41	60 88 99 10 041-6	61	60 88 99 69 101-9	72	60 88 99 09 002-1
51	60 88 99 69 051-5	62	60 88 99 69 102-7		

TYPE I0 — CORRIDOR FIRST

Built: 1974. Stainless steel TEE stock.
Bogies: Y 28.
Accommodation: 45F 2L 1W.
Heating: Air conditioned.
Computer Numbers: 61 88 1889 991–996/926–932.
Builder: .
Length over Buffers: 25.5 m.
Weight: 44 tonnes.
Max. Speed: 160 km/h.

131	133	135	141	143	145	146	147
132	134	136	142	144			

TYPE I0 — CORRIDOR SECOND

Built: 1974. Stainless steel TEE stock. Now declassiified.
Builder: .
Bogies: Y 28.
Accommodation: 45S 2L 1W.
Heating: Air conditioned.
Computer Numbers: 61 88 29 89 980–983.
Length over Buffers: 25.5 m.
Weight: 44 tonnes.
Max. Speed: 160 km/h.

151	152	153	154

SPECIAL SALOONS

No details available.
Computer Numbers: 60 88 99 80 001-5 & 60 88 99 89 002-4.

601	602

▲1930s vintage Class K1 open brake first No. 28120 at Houyet on 31/05/89 in the formation of the 09.58 ex-Namur. Note the 'birdcage' on the roof, the hopper ventilators and double slam doors. Pensylvania bogies are fitted. *Peter Fox*

▼Class K3 open second No. 22499 at Dinant on 31/05/89 in the same train as above. These coaches are of 1956 design and have drop windows and Schlieren bogies. *Peter Fox*

WAGONS LITS RESTAURANT CAR

Built: 1940–43.
Bogies: Pensylvania.
Accommodation: 48U.
Heating: Coal with through electric wiring.
Max. Speed: 140 km/h.
Computer Numbers: 51 88 88 50 007/11–13/74/85/17 & 61 88 88 70 265–269.
Builder: .
Length over Buffers: 23.45 m.
Weight: 52.6 tonnes.

4203 4212 | 4213 4230 | 4243 4247 | 4265 | 4266 | 4267 | 4268 | 4269

WAGONS LITS SLEEPING CAR TYPE MU

Built: 1967.
Bogies: Minden-Deutz M4.
Accommodation: .
Heating: Air conditioned.
Computer Numbers: 71 88 70 615–619.
Builder: .
Length over Buffers: m.
Weight: 61 tonnes.
Max. Speed: 160 km/h.

4792 | 4793 | 4794 | 4795 | 4796

WAGONS LITS SLEEPING CAR TYPE T2

Built: 1974–75.
Bogies: Minden-Deutz M6.
Accommodation: .
Heating: Air conditioned.
Computer Numbers: 71 88 75 70 156–161.
Builder: .
Length over Buffers: m.
Weight: 61 tonnes.
Max. Speed: 160 km/h.

5108 | 5109 | 5110 | 5151 | 5152 | 5153

1XXXX SERIES INTERNATIONAL STOCK

TYPE I4 CORRIDOR FIRST

Built: 1966.
Bogies: Schlieren.
Accommodation: 54F 2L 1W.
Heating: Electric.
Computer Numbers: 51 88 1970 402–419.
Builder: .
Length over Buffers: m.
Weight: 49 tonnes.
Max. Speed: 160 km/h.

b Formerly used on Benelux push-pull service.

11301 b 11302 b | 11303 b 11304 b | 11305 b 11306 b | 11307 b | 11308 b | 11315 | 11317 | 11318

TYPE I6 CORRIDOR FIRST

Built: 1977.
Bogies: Fiat Y0270.S.
Accommodation: 54F 2L.
Heating: Electric.
Computer Numbers: 61 88 19 70 601–620.
Builder: BN.
Length over Buffers: 26.40 m.
Weight: tonnes.
Max. Speed: 160 km/h.

11601 11602 11603 | 11604 11605 11606 | 11607 11608 11609 | 11610 11611 11612 | 11613 11614 | 11615 11616 | 11617 11618 | 11619 11620

TYPE I10 OPEN FIRST

Built: 1988.
Bogies: .
Accommodation: 66F 2L.
Heating: Electric .
Computer Numbers: 51 88 11 70 001–015.
Builder: BN.
Length over Buffers: 26.40 m.
Weight: 41 tonnes.
Max. Speed: 160/200 km/h.

11701	11703	11705	11707	11709	11711	11713	11715
11702	11704	11706	11708	11710	11712	11714	

TYPE I2 — SPECIAL TOUR COACH

Built: 1952. Converted from corridor first.
Builder: .
Bogies: Schlieren 22.
Accommodation: 1L & shower.
Heating: Electric.
Computer Numbers: 51 88 84 80 201/8/10/1.
Length over Buffers: 23.3 m.
Weight: 36 tonnes.
Max. Speed: 160 km/h.

11901	11908	11910	11911

TYPE I2 — CORRIDOR SECOND

Built: 1952. These vehicles have eight seats to a compartment.
Builder: .
Bogies: Schlieren 22.
Accommodation: 88S 2L.
Heating: Dual.
Computer Numbers: 51 88 21 80 203–212.
Length over Buffers: 23.30 m.
Weight: tonnes.
Max. Speed: 150 km/h.

12103	12105	12107	12108	12109	12110	12111	12112
12104	12106						

TYPE I2 — CORRIDOR SECOND

Built: 1952. Declassified firsts.
Bogies: Schlieren 22.
Accommodation: 64S 2L.
Heating: Dual.
Computer Numbers: 51 88 28 80 201–226.
Builder: .
Length over Buffers: 23.30 m.
Weight: tonnes.
Max. Speed: 150 km/h.

12151	12155	12159	12162	12165	12168	12171	12174
12152	12156	12160	12163	12166	12169	12172	12175
12153	12157	12161	12164	12167	12170	12173	12176
12154	12158						

TYPE I4 — CORRIDOR SECOND

Built: 1961/66.
Bogies: Schlieren 24 or 25.
Accommodation: 54S 2L.
Heating: Electric.
Builder: .
Length over Buffers: various.
Weight: tonnes.
Max. Speed: 160 km/h.
Computer Numbers: 51 88 29 70 401/2, 29 80 403/4, 29 70 407/9, 29 80 410–4/6.

12301	12303	12305	12307	12309	12311	12313	12315
12302	12304	12306	12308	12310	12312	12314	12316

TYPE I6 — CORRIDOR SECOND

Built: 1977–78.
Bogies: Fiat Y0270.S.
Accommodation: 66S 2L 1W.
Heating: Electric.
Computer Numbers: 61 88 21 70 601–660.
Builder: BN.
Length over Buffers: m.
Weight: 43 tonnes.
Max. Speed: 160 km/h.

Note: Some of these vehicles are being converted to couchettes and numbered in the 146XX series.

12601	12606	12611	12617	12622	12627	12632	12637
12602	12607	12613	12618	12623	12628	12633	12638
12603	12608	12614	12619	12624	12629	12634	12640
12604	12609	12615	12620	12625	12630	12635	12641
12605	12610	12616	12621	12626	12631	12636	12642

12643	12646	12649	12651	12653	12655	12657	12659
12644	12647	12650	12652	12654	12656	12658	12660
12645	12648						

TYPE I10 OPEN SECOND

Built: 1987–88.
Builder: BN.
Bogies: .
Length over Buffers: 26.40 m.
Accommodation: 86S 2L.
Max. Speed: 160/200 km/h.
Weight: 41 tonnes. (42.5 tonnes 12746–80).
Heating: Electric. 12746–80 have air conditioning.
Computer Numbers: 51 88 21–70 001–080.

12701	12711	12721	12731	12741	12751	12761	12771
12702	12712	12722	12732	12742	12752	12762	12772
12703	12713	12723	12733	12743	12753	12763	12773
12704	12714	12724	12734	12744	12754	12764	12774
12705	12715	12725	12735	12745	12755	12765	12775
12706	12716	12726	12736	12746	12756	12766	12776
12707	12717	12727	12737	12747	12757	12767	12777
12708	12718	12728	12738	12748	12758	12768	12778
12709	12719	12729	12739	12749	12759	12769	12779
12710	12720	12730	12740	12750	12760	12770	12780

TYPE I4 CORRIDOR COMPOSITE

Built: 1966.
Builder: BN.
Bogies: Schlieren 25.
Length over Buffers: 25.775 m.
Accommodation: 24F 36S 2L 2W.
Weight: 44 tonnes.
Heating: Electric.
Max. Speed: 160 km/h.
Computer Numbers: 50 88 39 70 403–410 & 50 88 30 70 411–417.
b Formerly used on Benelux push-pull service.

13202 b	13205 b	13208 b	13211	13213	13215	13217	13219
13203 b	13206 b	13209	13212	13214	13216	13218	13220
13204 b	13207 b	13210					

TYPE I3 SECOND COUCHETTE

Built: 1960.
Builder: .
Bogies: Schlieren 24.
Length over Buffers: 23.344 m.
Accommodation: 72S (54 berths) 2L 1W.
Weight: 45 tonnes.
Heating: Dual.
Max. Speed: 160 km/h.
Computer Numbers: 51 88 59 80 301–340.

14001	14006	14011	14016	14021	14026	14031	14036
14002	14007	14012	14017	14022	14027	14032	14037
14003	14008	14013	14018	14023	14028	14033	14038
14004	14009	14014	14019	14024	14029	14034	14039
14005	14010	14015	14020	14025	14030	14035	14040

TYPE I5 SECOND COUCHETTE

Built: 1967.
Builder: .
Bogies: Schlieren 25.
Length over Buffers: 26.40 m.
Accommodation: 60S (60 berths) 2L 3W.
Weight: 47 tonnes.
Heating: Dual.
Max. Speed: 160 km/h.
Computer Numbers: 51 88 50 70 501–545.

14101	14107	14113	14119	14125	14131	14136	14141
14102	14108	14114	14120	14126	14132	14137	14142
14103	14109	14115	14121	14127	14133	14138	14143
14104	14110	14116	14122	14128	14134	14139	14144
14105	14111	14117	14123	14129	14135	14140	14145
14106	14112	14118	14124	14130			

TYPE ID BRAKE VAN

Built: 1939.
Bogies: Pensylvania.
Accommodation: 1L.
Heating: Dual.
Computer Numbers: 51 88 95 40 914–921.
Builder: .
Length over Buffers: 16.16 m.
Weight: 37 tonnes.
Max. Speed: 140 km/h.

17101 | 17102 | 17103 | 17104 | 17105 | 17106 | 17107 | 17108

TYPE ID BRAKE VAN

Built: 1978.
Bogies: Fiat YO 332.
Accommodation: .
Heating: Electric.
Computer Numbers: 51 88 95 70 901–934.
Builder: .
Length over Buffers: 26.40 m.
Weight: 39 tonnes.
Max. Speed: 160 km/h.

17401 | 17402 | 17403 | 17404 | 17405 | 17406 | 17407 | 17408 | 17409 | 17410 | 17411 | 17412 | 17413 | 17414 | 17415 | 17416 | 17417 | 17418 | 17419 | 17420 | 17421 | 17422 | 17423 | 17424 | 17425 | 17426 | 17427 | 17428 | 17429 | 17430 | 17431 | 17432 | 17433 | 17434

BRAKE VAN

No details except that 17815 is a generator.
Computer Numbers: 60 88 99 40 021–035.

17801 | 17802 | 17803 | 17804 | 17805 | 17806 | 17807 | 17808 | 17809 | 17810 | 17815

TYPE I1 BAR

Built: 1933. Ex 12248.
Bogies: Schlieren 27.
Accommodation: Nil.
Heating: Electric with through steam pipe.
Computer Number: 51 88 89 80 001-8.
Builder: .
Length over Buffers: 22.30 m.
Weight: 45 tonnes.
Max. Speed: 160 km/h.

17901

TYPE I6 BAR

Built: 1933.
Bogies: Fiat 31.
Accommodation: Nil.
Heating: Electric.
Computer Number: 61 88 89 70 002-6.
Builder: .
Length over Buffers: 26.40 m.
Weight: 48 tonnes.
Max. Speed: 160 km/h.

17902

2XXXX SERIES TYPE K STOCK

TYPE K1 OPEN FIRST

Built: 1933–34.
Bogies: Pensylvania.
Accommodation: 70F 1L.
Heating: Electric.
Builder: .
Length over Buffers: 23.32 m.
Weight: 43.1 tonnes.
Max. Speed: 140 km/h.
Computer Numbers: 50 88 18 40 001–123. Some of those shown now withdrawn.

21001 | 21002 | 21005 | 21007 | 21008 | 21012 | 21015 | 21016 | 21017 | 21023 | 21024 | 21026 | 21027 | 21028 | 21029 | 21030 | 21033 | 21036 | 21037 | 21038 | 21123

TYPE K2 OPEN SECOND

Built: 1953.
Builder: .
Bogies: Schlieren or Pensylvania.
Length over Buffers: 23.32 m.
Accommodation: 108S 1L.
Weight: 32.8 tonnes.
Heating: Dual.
Max. Speed: 140 km/h.
Computer Numbers: 50 88 21 48 155-5, 159-7, 163-9.

22205 | 22209 | 22213

TYPE K3 OPEN SECOND

Built: 1956–58.
Builder: BN.
Bogies: Schlieren.
Length over Buffers: 23.22 m.
Accommodation: 108S 1L.
Weight: 35.7 tonnes.
Heating: Dual.
Max. Speed: 140 km/h.
Computer Numbers: 50 88 21 48 300–399. (Note: last digit one less than SNCB number.)

22401	22414	22427	22439	22451	22465	22477	22489
22402	22415	22428	22440	22452	22466	22478	22490
22403	22416	22429	22441	22453	22467	22479	22491
22404	22417	22430	22442	22455	22468	22480	22492
22405	22418	22431	22443	22456	22469	22481	22493
22406	22419	22432	22444	22457	22470	22482	22494
22407	22420	22433	22445	22458	22471	22483	22495
22408	22421	22434	22446	22459	22472	22484	22496
22409	22422	22435	22447	22460	22473	22485	22497
22410	22423	22436	22448	22461	22474	22486	22498
22411	22424	22437	22449	22462	22475	22487	22499
22412	22425	22438	22450	22464	22476	22488	22500
22413	22426						

TYPE K1 OPEN COMPOSITE

Built: 1933.
Builder: .
Bogies: Pensylvania.
Length over Buffers: 23.32 m.
Accommodation: 31F 39S 1L.
Weight: 41.5 tonnes.
Heating: Electric.
Max. Speed: 140 km/h.
Computer Number: 50 88 39 48 001-1.

23001

TYPE K1 OPEN BRAKE FIRST

Built: 1933–35. These vintage vehicles are of 'birdcage' design, with a guard's lookout on the roof.
Builder: .
Bogies: Pensylvania.
Length over Buffers: 23.32 m.
Accommodation: 50F 1L.
Weight: 41.5 tonnes.
Heating: Dual.
Max. Speed: 140 km/h.
Computer Numbers: 50 88 81 48 005–025.

28105	28107	28110	28115	28117	28119	28120	28127
28106	28108	28112	28116	28118			

TYPE K1 OPEN BRAKE FIRST

Built: 1933–34.
Builder: .
Bogies: Pensylvania.
Length over Buffers: 23.32 m.
Accommodation: 39F 1L.
Weight: 43 tonnes.
Heating: Dual.
Max. Speed: 140 km/h.
Computer Numbers: 50 88 87 48 001–015.

28901	28903	28905	28907	28909	28911	28913	28915
28902	28904	28906	28908	28910	28912	28914	

TYPE K2 BRAKE SECOND

Built: 1953.
Bogies: Pensylvania.
Accommodation: 69S 1L.
Heating: Dual.
Computer Number: 50 88 82 48 154-4.
Builder: .
Length over Buffers: 23.32 m.
Weight: 32.7 tonnes.
Max. Speed: 140 km/h.

29204

4XXXX SERIES TYPE M1, M2 & M3 STOCK

TYPE M3 OPEN BRAKE COMPOSITE

Built: 1960–62.
Bogies: Improved Pensylvania.
Accommodation: 44F 29S 1L.
Heating: Steam.
Computer Numbers: 50 88 81 26 951–960.
Builder: .
Length over Buffers: 24.00 m.
Weight: 30 tonnes.
Max. Speed: 120 km/h.

40001	40003	40004	40005	40006	40007	40008	40010
40002							

TYPE M2 OPEN FIRST

Built: 1958–60.
Bogies: Schlieren type 23.
Accommodation: 68F 1L.
Heating: Dual.
Computer Numbers: 50 88 18 48 (38 e) 601–635.
Builder: .
Length over Buffers: 24.00 m.
Weight: 34.5 tonnes.
Max. Speed: 120 km/h.

e Electric heating only.

41001 e	41006	41011	41016	41020	41024	41028	41032
41002	41007	41012 e	41017	41021	41025	41029	41033
41003 e	41008	41013	41018	41022	41026	41030	41034
41004	41009	41014	41019	41023	41027	41031 e	41035
41005	41010	41015					

TYPE M1 OPEN SECOND

Built: 1937. Push-pull fitted.
Bogies: Pensylvania.
Accommodation: 94S 1L.
Heating: Steam.
Computer Numbers: 50 88 27 26 602/5/30/43/44/63.
Builder: .
Length over Buffers: 22.76 m.
Weight: 43.9 tonnes.
Max. Speed: 120 km/h.

42002	42014	42062	42094	42097	42128

TYPE M2 OPEN SECOND

Built: 1958–60.
Bogies: Schlieren type 23.
Accommodation: 106S 1L.
Heating: Dual.
Computer Numbers: 50 88 20 48 (38 e) 601–950.
Builder: .
Length over Buffers: 24.00 m.
Weight: 33.6 tonnes.
Max. Speed: 120 km/h.

e Electric heating only.
p Push-pull fitted.

42301	42308	42315	42322	42329	42336	42343	42350
42302	42309 e	42316	42323	42330	42337	42344	42351
42303 e	42310	42317	42324	42331	42338	42345	42352
42304	42311 p	42318	42325	42332	42339	42346	42353
42305	42312	42319	42326	42333	42340	42347 p	42354
42306	42313	42320 e	42327	42334	42341	42348 p	42355 e
42307	42314 e	42321	42328	42335	42342	42349 p	42356

▲High platform unit 034 in the old green livery is seen leaving Antwerpen Centraal with the 15.02 stopping service to Brussels Midi on 01/06/89. *Peter Fox*

▼A Sabena unit, No. 595 in the special white & blue livery of this class at Brussels Nord with the 13.19 Brussels Airport–Brussels Centraal on 01/06/89. *Peter Fox*

INTER-CITY EMUs

▲'Break' unit No. 353 on a Herstal–Tournai train at Liège Jonfosse on 16/05/85. *Graham Scott-Lowe*

▼Orange and grey liveried four-car unit No. 834 approaches Brussels on 30/05/89 with a service from Antwerpen. A similar unit in the new red livery can be seen coupled on the rear. *Peter Fox*

▲It is easy to see from the front-end design why the new Class AM86 EMUs have gained the nickname of 'Snorkels'. 927 is pictured outside the La Brougoise–Nivelles works at Brugge on 22/05/89. *David Beynon*

▼A new batch of postal units has been produced by conversion of units in the 051–128 series. They have roller shutter doors and a distinctive red livery as shown by this photograph of 967 at Brussels Midi on 27/05/89. *Graham Scott-Lowe*

DUAL-VOLTAGE ELECTRICS

▲1500/3000 V dc Class 11 No. 1181 is seen piloting triple voltage (1500/3000 V dc, 25 kV ac) Class 15 No. 1503 on an Amsterdam–Ventimiglia working at Roosendaal on 17/04/86. When this photograph was taken, the Class 11 was being run in. The Class 15s arenow banned from the Netherlands because they are not fitted with NS ATB, and their workings are now in the heads of the Class 25.5, displaced from the previous Benelux push-pull work. *David Beynon*

▼The Class 12s are also dual voltage, but in this case have a 25 kV ac capability for working into France. 1205 is seen on a commuter train of the new M5 double-decker stock at Brussels Nord on 22/05/89. *David Beynon*

▲The quadruple-voltage Class 18s can, in theory, work anywhere in Europe. They are typically French in appearance, and are in fact identical to the SNCF CC 40100 Class. No. 1802 in the grey/blue/yellow and silver livery makes a fine sight at Liège Guillemins on 30/05/89 as it leaves with the Eurocity Parsifal, the 07.25 Paris Nord–Köln, formed of SNCF Corail stock. *Peter Fox*

▼Class 21 No. 2109 is painted in a special advertising livery and was photographed on an exhibition train at the other end of Liège Guillemins on 03/05/89. *John Armistead*

OLDER ELECTRICS

▲Class 22 No. 2202 was one of the first to be painted in the now-superseded blue & yellow livery when photographed on 28/07/79 ex-works at Ronet. It still carries this livery. *Brian Garvin*

▼Class 23 No. 2346 is still in the even older green livery. It is seen at Brussels Nord with the 17.02 to Chatelet formed of M4 stock. *David Beynon*

▲Class 45 diesel railcar No. 4504 at Virton forming the 12.15 to Bastogne Sud on 04/05/89. This service was extended to Bastogne Nord as from 28/05/89. All railcars are painted in this red and yellow livery. *John Armistead*

▼Class 54 diesel loco No. 5407 in old livery runs light through Namur on 02/05/89. *John Armistead*

DIESELS ON PASSENGER & FREIGHT

▲Class 55 No. 5525 at Jemelle on the 16.00 to Liège on 04/05/89 with a rake of green M2 stock. A similar loco can be seen at the other end of the train having worked the incoming service from Liège.
John Armistead

▼Class 73 No. 7339 is seen on a freight trip at Liège Guillemins on 09/04/82. The shunter liveries are the only ones to have remained stable during the last ten years. *Brian Garvin*

MODERN BELGIAN COACHING STOCK

▲International stock is now painted in the ubiquitous European standard orange with a grey stripe. Type I10 open second No. 12719 can be seen next to type I4 composite on the 08.10 Köln–Oostende. The unofficial Belgian number can be clearly seen crossed out at the bottom left-hand side of the coach. *Peter Fox*

▼A rake of M4 stock at Liège Guillemins with open first No. 51014 nearest the camera. The doors on these vehicles are at the ends of the coaches, whereas on the second Class vehicles they divide the saloon into three parts. *Peter Fox*

ELECTRIC-HAULED PASSENGER

▲1100 Class No. 1139 passes through Barendrecht on 29/05/89 with the 12.12 Den Haag–Köln formed of DB stock with an NS BKD at the rear. *Peter Fox*

▼1200 Class Co–Co No. 1202 at Amsterdam CS with the 14.31 Heerlen–Zandvoort-an-See train on 29/05/89. *Ian Futers*

▲1300 Class No. 1312 with DB stock has arrived at Amsterdam CS with an Intercity working from Köln on 28/05/89. *Ian Futers*

▼The Amsterdam–Brussels service is worked by Belgian Class 11 locos with NS ICR push-pull sets in a distinctive yellow and Bordeaux red livery. The 11.26 ex Amsterdam is seen approaching Rotterdam CS on 29/05/89 with driving trailer No. 113 leading. The third vehicle is new ICR4 open second No. 583 in NS Intercity livery, one of 5 such vehicles fitted with push-pull jumpers for Benelux reserve use. *Peter Fox*

DIESELS OLD & NEW

◄There are now only two diese locomotives in the old NS red/browi livery. One of these, 2200 Class Nc 2275 was pictured at Maastricht on 29 05/89. *Peter Fo*

▼The new 6400 Class thyristor-controlled three-phase drive Bo–Bos went into revenue earning service with the commencement of the 1989 summer timetable. 6401 'Mijndert' is seen shunting at Delfzijl on 02/06/89.

Peter Fox

NEW MULTIPLE UNITS

▲IC3 EMUs Nos. 4049 & 4031 at Groningen on 02/06/89 with the 12.37 to Hoofddorp. *Peter Fox*

▼3200 Class 'Wadloper' DMU No. 3207 at Groningen on 02/06/89 having arrived on the 1147 ex-Delfzijl. *Peter Fox*

CFL CONTRASTS

▲At Trois Vierges, 1800 Class Co–Co No. 1819 'PRINCE HENRI' is seen having arrived with the 17.25 ex-Luxembourg with DMU No. 201 to the right. *Peter Fox*

▼EMU No. 253 of typically French design is seen at Bettembourg on 26/05/89. *Graham Scott-Lowe*

▲Streamlined inside cylinder Atlantic No. 12.004 emerges from a rock tunnel on 25/08/85 at the approach to Dinant in the Meuse Valley with an enthusiasts' special from Namur. *David Rodgers*

▼2–8–0 No. 29.013 on exhibition and acting as standby for 12.004 on the same day at Dinant. It had earlier double-headed with the Atlantic from Namur. *David Rodgers*

THE RETURN OF ELECTRA

▲BR No. 27000 'ELECTRA' became NS No. 1502 in 1970. Withdrawn in 1986, the loco has been restored by the EM2 Society and is pictured on its return to the Netherlands at Utrecht on 17/06/89. *Charles Petty*

42357	42395	42432	42469	42507	42543 e	42579	42615
42359	42396	42433 p	42470	42508	42544	42580	42616
42360	42397	42434 e	42471	42509	42545	42581	42617
42361	42398	42435 e	42472	42510	42546	42582	42618
42362	42399	42436	42473	42511 e	42547	42583	42619
42363 e	42400	42437	42474	42512	42548	42584	42620 p
42364 p	42401 e	42438	42475	42513	42549 p	42585	42621
42365	42402	42439	42476	42514	42550	42586 p	42622 p
42366	42403 p	42440	42477	42515	42551	42587	42623 e
42367	42404	42441	42478	42516	42552	42588 p	42624
42368	42405	42442	42479	42517 e	42553	42589	42625
42369	42406	42443	42480	42518	42554	42590 p	42626 e
42370 p	42407	42444	42482	42519	42555	42591	42627
42371	42408	42445 e	42483 e	42520 p	42556	42592	42628 p
42372	42409	42446 e	42484	42521 e	42557	42593 e	42629
42373	42410	42447 e	42485	42522	42558 p	42594	42630
42374 e	42411	42448	42486	42523	42559	42595 e	42631 p
42375 p	42412	42449	42487	42524	42560	42596 e	42632 e
42376	42413	42450 e	42488	42525 p	42561	42597 e	42633 e
42377	42414	42451	42489	42526 e	42562	42598	42634 e
42378 p	42415	42452	42490	42527	42563	42599	42635 e
42379	42416	42453	42491	42528	42564 p	42600	42636
42380	42417 p	42454	42492	42529	42565 e	42601 e	42637
42381 e	42418	42455	42493	42530 e	42566	42602	42638 p
42382 e	42419 e	42456	42494 p	42531 p	42567	42603	42639
42383	42420 p	42457	42495	42532	42568	42604	42640
42384	42421	42458	42496 p	42533	42569	42605	42641 e
42385	42422 e	42459 p	42497	42534	42570	42606	42642 p
42386	42423 e	42460	42498	42535 e	42571 p	42607	42643
42387	42424 e	42461	42499	42536 p	42572	42608 e	42644 p
42388	42425	42462 e	42500 p	42537 p	42573 e	42609 e	42645 p
42389 e	42426	42463	42501	42538	42574 p	42610	42646
42390	42427	42464	42502 p	42539	42575	42611	42647 p
42391	42428	42465	42503	42540	42576 p	42612 e	42648
42392 e	42429	42466	42504	42541	42577 p	42613	42649 p
42393	42430	42467	42505	42542 p	42578	42614	42650 e
42394 p	42431	42468	42506				

TYPE M3 — OPEN SECOND

Built: 1960–62.
Bogies: Improved Pensylvania.
Accommodation: 114S 1L.
Heating: Steam.
Computer Numbers: 50 88 21 26 951–986.
Builder: .
Length over Buffers: 24.00 m.
Weight: 30 tonnes.
Max. Speed: 120 km/h.

42801	42806	42811	42816	42821	42825	42829	42833
42802	42807	42812	42817	42822	42826	42830	42834
42803	42808	42813	42818	42823	42827	42831	42835
42804	42809	42814	42819	42824	42828	42832	42836
42805	42810	42815	42820				

TYPE M1 — OPEN COMPOSITE

Built: 1937. Push-pull fitted.
Bogies: Pensylvania.
Accommodation: 38F 36S 1L.
Heating: Steam.
Computer Numbers: 50 88 38 26 624-7 and 637-9.
Builder: .
Length over Buffers: 22.76 m.
Weight: 42.2 tonnes.
Max. Speed: 120 km/h.

43045	43078

TYPE M2 — OPEN COMPOSITE

Built: 1958–60.
Bogies: Schlieren type 23.
Builder: .
Length over Buffers: 24.00 m.

Accommodation: 36F 47S 1L.
Heating: Dual.
Computer Numbers: 50 88 39 48 (38 e) 601–704.
Weight: 34.4 tonnes.
Max. Speed: 120 km/h.

e Electric heating only.
p Push-pull fitted.

43201	43214 p	43227	43240 e	43253	43266	43279	43292
43202 e	43215 p	43228	43241	43254	43267 e	43280 e	43293
43203	43216 e	43229 p	43242	43255	43268	43281 p	43294
43204 e	43217 p	43230 p	43243	43256 p	43269	43282 p	43295 p
43205	43218 p	43231	43244 e	43257 e	43270 e	43283	43297 p
43206	43219	43232	43245	43258	43271	43284	43298 p
43207	43220	43233 e	43246	43259	43272 e	43285 p	43299 e
43208	43221	43234 p	43247 e	43260 e	43273	43286 p	43300
43209	43222 p	43235	43248	43261	43274	43287	43301
43210	43223	43236	43249	43262	43275 p	43288	43302
43211	43224	43237	43250	43263 e	43276	43289	43303 p
43212	43225	43238	43251 e	43264	43277	43290 e	43304
43213 p	43226	43239	43252 p	43265	43278	43291 p	

TYPE M1 DRIVING BRAKE OPEN SECOND

Built: 1937. Push-pull driving trailer.
Bogies: Pensylvania.
Accommodation: 77S 1L.
Heating: Steam.
Computer Numbers: 50 88 82 26 649-9 and 658-0.
Builder: .
Length over Buffers: 22.76 m.
Weight: 41.9 tonnes.
Max. Speed: 120 km/h.

49108	49117

TYPE M2 BRAKE OPEN SECOND

Built: 1958–60.
Bogies: Schlieren type 23.
Accommodation: 75S 1L.
Heating: Dual.
Computer Numbers: 50 88 82 48 (38 e) 601–716.
Builder: .
Length over Buffers: 24.00 m.
Weight: 32.4 tonnes.
Max. Speed: 120 km/h.

e Electric heating only.
p Converted to push-pull driving trailer.

49201	49216	49233	49247	49261	49275	49289	49303
49202	49217	49234 p	49248	49262	49276	49290	49304
49203	49218 e	49235	49249	49263	49277	49291	49305 p
49204	49219 e	49236 p	49250	49264 p	49278	49292 p	49306
49205	49222 e	49237 p	49251	49265	49279	49293 p	49307 p
49206	49223	49238	49252	49266 p	49280 p	49294	49308
49207	49224	49239	49253 e	49267	49281	49295 e	49309
49208	49225	49240	49254	49268	49282	49296	49310
49209	49226	49241 p	49255	49269	49283	49297	49311
49210 p	49227	49242	49256 e	49270	49284 p	49298 p	49312
49211 p	49228	49243	49257	49271 p	49285 p	49299	49313
49212 p	49229	49244	49258 e	49272	49286 p	49300	49314
49213	49230	49245	49259	49273	49287	49301	49315
49214	49231	49246	49260 e	49274 p	49288	49302	49316 p
49215	49232 p						

TYPE M2 DRIVING BRAKE OPEN SECOND

Built: 1958–60. Push-pull driving trailer.
Bogies: Schlieren type 23.
Accommodation: 65S 1L.
Heating: Dual.
Computer Numbers: 50 88 87 48 (38 e) 601–615.
Builder: .
Length over Buffers: 24.00 m.
Weight: 32.4 tonnes.
Max. Speed: 120 km/h.

e Electric heating only.

49901	49903	49905	49907 e	49909	49911	49913	49915
49902	49904 e	49906	49908 e	49910	49912	49914	

5XXXX SERIES TYPE M4 & M5 STOCK

TYPE M4 — OPEN FIRST

Built: 1979–80.
Bogies: Fiat type Y32.
Accommodation: 80F 1L.
Heating: Electric.
Computer Numbers: 50 88 19 78 001–050.
Builder: .
Length over Buffers: 24.26 m.
Weight: 38 tonnes.
Max. Speed: 160 km/h.

p Push-pull fitted.

51001	51008	51015	51021	51027	51033	51039 p	51045 p
51002	51009	51016	51022	51028	51034 p	51040 p	51046 p
51003	51010	51017	51023	51029	51035 p	51041 p	51047 p
51004	51011	51018	51024	51030	51036 p	51042 p	51048 p
51005	51012	51019	51025	51031	51037 p	51043 p	51049 p
51006	51013	51020	51026	51032	51038 p	51044 p	51050 p
51007	51014						

TYPE M5 — DOUBLE-DECK OPEN FIRST

Built: 1988–89. Push-pull fitted.
Bogies: .
Accommodation: 142F 2L.
Heating: Electric.
Computer Numbers: 50 88 16 38 001–015.
Builder: .
Length over Buffers: . m.
Weight: 44 tonnes.
Max. Speed: 140 km/h.

51501	51503	51505	51507	51509	51511	51513	51515
51502	51504	51506	51508	51510	51512	51514	

TYPE M4 — OPEN SECOND

Built: 1980–83.
Bogies: Fiat type Y32.
Accommodation: 112S 1L.
Heating: Electric.
Computer Numbers: 50 88 20 78 001–430.
Builder: .
Length over Buffers: 24.26 m.
Weight: 39 tonnes.
Max. Speed: 160 km/h.

p Push-pull fitted.

52001	52024	52047	52070	52093	52116	52139	52162
52002	52025	52048	52071	52094	52117	52140	52163
52003	52026	52049	52072	52095	52118	52141	52164
52004	52027	52050	52073	52096	52119	52142	52165
52005	52028	52051	52074	52097	52120	52143	52166
52006	52029	52052	52075	52098	52121	52144	52167
52007	52030	52053	52076	52099	52122	52145	52168
52008	52031	52054	52077	52100	52123	52146	52169
52009	52032	52055	52078	52101	52124	52147	52170
52010	52033	52056	52079	52102	52125	52148	52171
52011	52034	52057	52080	52103	52126	52149	52172
52012	52035	52058	52081	52104	52127	52150	52173
52013	52036	52059	52082	52105	52128	52151	52174
52014	52037	52060	52083	52106	52129	52152	52175
52015	52038	52061	52084	52107	52130	52153	52176
52016	52039	52062	52085	52108	52131	52154	52177
52017	52040	52063	52086	52109	52132	52155	52178
52018	52041	52064	52087	52110	52133	52156	52179
52019	52042	52065	52088	52111	52134	52157	52180
52020	52043	52066	52089	52112	52135	52158	52181
52021	52044	52067	52090	52113	52136	52159	52182
52022	52045	52068	52091	52114	52137	52160	52183
52023	52046	52069	52092	52115	52138	52161	52184

52185	52216	52247	52278 p	52309 p	52340 p	52371 p	52401 p
52186	52217	52248 p	52279 p	52310 p	52341 p	52372 p	52402 p
52187	52218	52249 p	52280 p	52311 p	52342 p	52373 p	52403 p
52188	52219	52250 p	52281 p	52312 p	52343 p	52374 p	52404 p
52189	52220	52251 p	52282 p	52313 p	52344 p	52375 p	52405 p
52190	52221	52252 p	52283 p	52314 p	52345 p	52376 p	52406 p
52191	52222	52253 p	52284 p	52315 p	52346 p	52377 p	52407 p
52192	52223	52254 p	52285 p	52316 p	52347 p	52378 p	52408 p
52193	52224	52255 p	52286 p	52317 p	52348 p	52379 p	52409 p
52194	52225	52256 p	52287 p	52318 p	52349 p	52380 p	52410 p
52195	52226	52257 p	52288 p	52319 p	52350 p	52381 p	52411 p
52196	52227	52258 p	52289 p	52320 p	52351 p	52382 p	52412 p
52197	52228	52259 p	52290 p	52321 p	52352 p	52383 p	52413 p
52198	52229	52260 p	52291 p	52322 p	52353 p	52384 p	52414 p
52199	52230	52261 p	52292 p	52323 p	52354 p	52385 p	52415 p
52200	52231	52262 p	52293 p	52324	52355 p	52386 p	52416 p
52201	52232	52263 p	52294 p	52325 p	52356 p	52387 p	52417 p
52202	52233	52264 p	52295 p	52326 p	52357 p	52388 p	52418 p
52203	52234	52265 p	52296 p	52327 p	52358 p	52389 p	52419 p
52204	52235	52266 p	52297 p	52328 p	52359 p	52390 p	52420 p
52205	52236	52267 p	52298 p	52329 p	52360 p	52391 p	52421 p
52206	52237	52268 p	52299 p	52330 p	52361 p	52392 p	52422 p
52207	52238	52269 p	52300 p	52331 p	52362 p	52393 p	52423 p
52208	52239	52270 p	52301 p	52332 p	52363 p	52394 p	52424 p
52209	52240	52271 p	52302 p	52333 p	52364 p	52395 p	52425 p
52210	52241	52272 p	52303 p	52334 p	52365 p	52396 p	52426 p
52211	52242	52273 p	52304 p	52335 p	52366 p	52397 p	52427 p
52212	52243	52274 p	52305 p	52336 p	52367 p	52398 p	52428 p
52213	52244	52275 p	52306 p	52337 p	52368 p	52399 p	52429 p
52214	52245	52276 p	52307 p	52338 p	52369 p	52400 p	52430 p
52215	52246	52277 p	52308 p	52339 p	52370 p		

TYPE M5 DOUBLE-DECK OPEN SECOND

Built: 1988–89. Push-pull fitted.
Bogies: .
Accommodation: 146S 2L.
Heating: Electric.
Computer Numbers: 50 88 16 38 001–015.

Builder: .
Length over Buffers: . m.
Weight: 43.8 tonnes.
Max. Speed: 140 km/h.

52501	52514	52526	52538	52550	52562	52574	52586
52502	52515	52527	52539	52551	52563	52575	52587
52503	52516	52528	52540	52552	52564	52576	52588
52504	52517	52529	52541	52553	52565	52577	52589
52505	52518	52530	52542	52554	52566	52578	52590
52506	52519	52531	52543	52555	52567	52579	52591
52507	52520	52532	52544	52556	52568	52580	52592
52508	52521	52533	52545	52557	52569	52581	52593
52509	52522	52534	52546	52558	52570	52582	52594
52510	52523	52535	52547	52559	52571	52583	52595
52511	52524	52536	52548	52560	52572	52584	52596
52512	52525	52537	52549	52561	52573	52585	52597
52513							

TYPE M4 BRAKE OPEN FIRST

Built: 1982.
Bogies: Fiat type Y32.
Accommodation: 56F 1L.
Heating: Electric.
Computer Numbers: 50 88 81 78 001–033.

Builder: .
Length over Buffers: 24.26 m.
Weight: 37.2 tonnes.
Max. Speed: 160 km/h.

58001	58005	58009	58013	58017	58021	58025	58029
58002	58006	58010	58014	58018	58022	58026	58030
58003	58007	58011	58015	58019	58023	58027	58031
58004	58008	58012	58016	58020	58024	58028	58032

58033

TYPE M4 DRIVING BRAKE OPEN FIRST

Built: 1983. Push-pull driving trailer.
Bogies: Fiat type Y32.
Accommodation: 48F 1L.
Heating: Electric.
Computer Numbers: 50 88 81 78 034–065.
Builder: .
Length over Buffers: 24.26 m.
Weight: 39 tonnes.
Max. Speed: 160 km/h.

58034	58038	58042	58046	58050	58054	58058	58062
58035	58039	58043	58047	58051	58055	58059	58063
58036	58040	58044	58048	58052	58056	58060	58064
58037	58041	58045	58049	58053	58057	58061	58065

TYPE M4 BRAKE OPEN SECOND

Built: 1983–84.
Bogies: Fiat type Y32.
Accommodation: 64S 1L.
Heating: Electric.
Computer Numbers: 50 88 87 78 001–035.
Builder: .
Length over Buffers: 24.26 m.
Weight: 39.5 tonnes.
Max. Speed: 160 km/h.

p Converted to push-pull driving trailer.

59901	59906	59911	59916	59920	59924	59928 p	59932 p
59902	59907	59912	59917	59921	59925	59929 p	59933 p
59903	59908	59913	59918	59922	59926 p	59930	59934 p
59904	59909	59914	59919	59923	59927 p	59931	59935
59905	59910	59915					

TYPE M5 DOUBLE-DECK DRIVING OPEN SECOND

Built: 1988–89. Push-pull driving trailer.
Bogies: .
Accommodation: 118S 2L.
Heating: Electric.
Computer Numbers: 50 88 82 38 001–035.
Builder: .
Length over Buffers: . m.
Weight: 49.3 tonnes.
Max. Speed: 140 km/h.

59951	59954	59957	59959	59961	59963	59965	59967
59952	59955	59958	59960	59962	59964	59966	59968
59953	59956						

POSTAL VANS

Built: 1933–34. Vehicles marked g have generators.
Bogies: Pensylvania.
Weight: 30.6 (33.6*, 46g) tonnes.
Max. Speed: 120 (140*) km/h.
Computer Numbers: 50 88 92 66 901–911 (50 88 99 49 301–304*).
Length over Buffers: 15.60 m.
Heating: Steam (Dual*).

77011	77013	77015	77020 g	77301 *	77302 *	77303 *	77304 *
77012	77014	77019 g	77023 g				

M2 Push-pull driving trailer No. 49292 is the leading vehicle of the 15.09 Neerpelt as it leaves Antwerpen Centraal on 01/06/89. *Peter Fox*

▲'Sik' No. 225 at Haarlem on 31/07/88. *Alex Dasi-Sutton*

▼Class 500 diesel electric shunter No. 628 shunts wagons in Roosendaal marshalling yard on 14/08/86. *David Brown*

2. NETHERLANDS RAILWAYS (NS)

The Dutch name for the Netherlands Railways is Nederlandse Spoorwegen, hereafter abreviated to NS. The NS is a relatively small system and approximately two thirds of all routes are electrified at 1500 V dc with overhead wire collection. Comprehensive maps of the system are to be found on the next four pages. Language presents no problem, since most Dutch people speak good English, but restaurant menus are often in dutch only, so a dictionary or phrase book can still be useful. Different numbering series are used for locos and multiple units.

DEPOTS & WORKSHOPS

Not all NS locos and units have depot allocations. The vast majority are common-user but there are some definite allocations, as all the electric locos are allocated to Maastricht whilst certain 2200s are definitely allocated to Feijenoord and the 6400s are at Zwolle. The situation can be summarised as follows:

Electric locos	Maastricht
Diesel locos	Feijenoord, Watergraafsmeer, Zwolle
EMUs	Amsterdam Zaanstraat, Leidschendamm, Maastricht
DMUs	Maastricht, Zwolle, Onnen

The main workshops are at Haarlem (for all units) and Tilburg (for all locos and the power units of DMUs.

There are many places where EMUs stable, but the main locations where units and some locos will be found are: Alkmaar, Amersfoort, Arnhem, Botlek*, Den Haag CS, Eindhoven, Goes, Groningen, Heerlen, Hengelo, Kijfhoek Yard*, Leeuwarden, Nijmegen, Roosendaal, Rotterdan CS, Sittard, Utrecht, Venlo, Vlissingen, Waalhaven Zuid* and Zutphen. (* – locos only).

PASSENGER TRAINS

The NS timetable is almost entirely regular interval with most routes having half-hourly services and many routes four trains per hour.
Most NS passenger trains are worked by EMUs and the types of trains worked by these are detailed in the individual class details. All long distance international trains are loco-hauled. Principal internal loco-hauled routes (including workings into adjacent countries are:

Amsterdam–Den Haag–Rotterdam–Dordrecht–Roosendaal–Brussels. These are push-pull trains with SNCB dual-voltage class 11 electric locos and plan ICR3 stock. Trains which go through to France are generally SNCB class 25.5 hauled.

Zandvoort–Haarlem–Amsterdam–Utrecht–Den Bosch–Eindhoven–Sittard–Maastricht/Heerlen. These trains are loco hauled with either class 1200 or 1600, but 1300s are also used. Because loads are heavy, the ICR coaches are strengthened with plan W stock and 1100s are rare.

Amsterdam–Rotterdam–Utrecht–Arnhem/Nijmegen–Köln/Hagen. Amsterdam–Köln/Hagen services are worked by 1100, 1300 and 1600 electrics with some 1200 turns. The Köln/Hagen services are worked forward from Emmerich by DB class 110 electrics. The Köln via Nijmegen services are worked forward from Arnhem with DB 215, 216 or 218 series diesel hydraulics. DB stock is generally used on these, including some trains which terminate at Arnhem!

Den Haag–Rotterdam–Dordrecht–Tilburg–Eindhoven–Venlo–(Köln). Den Haag–Köln services are loco hauled mainly with class 1100s, although other classes also appear. Some turns which terminate at Venlo are also loco-hauled. Because of a hauled stock shortage on the NS, trains are now formed mainly of DB stock instead of NS ICR stock. Köln services are usually worked forward from Venlo by DB 110 series electrics.

Table 60. Zwolle–Deventer–Arnhem–Nijmegen–Den Bosch–Tilburg–Roosendaal. This service is at present worked by 'Hondekop' EMUs, but because many of these units will be out of service due to a crash programme of blue asbestos removal, this service *will revert to loco-haulage* with the winter 1989 timetable with hired Belgian M2 stock! The regular users of this line are reported to be up in arms at the prospect!

Odd trains on other routes are also loco-hauled, particularly in peak hours. A peak hour working between Tilburg and Utrecht is also expected to receive Belgian stock, but this will get declassified firsts!

Nederlandse Spoorwegen

NS NETWORK MAPS

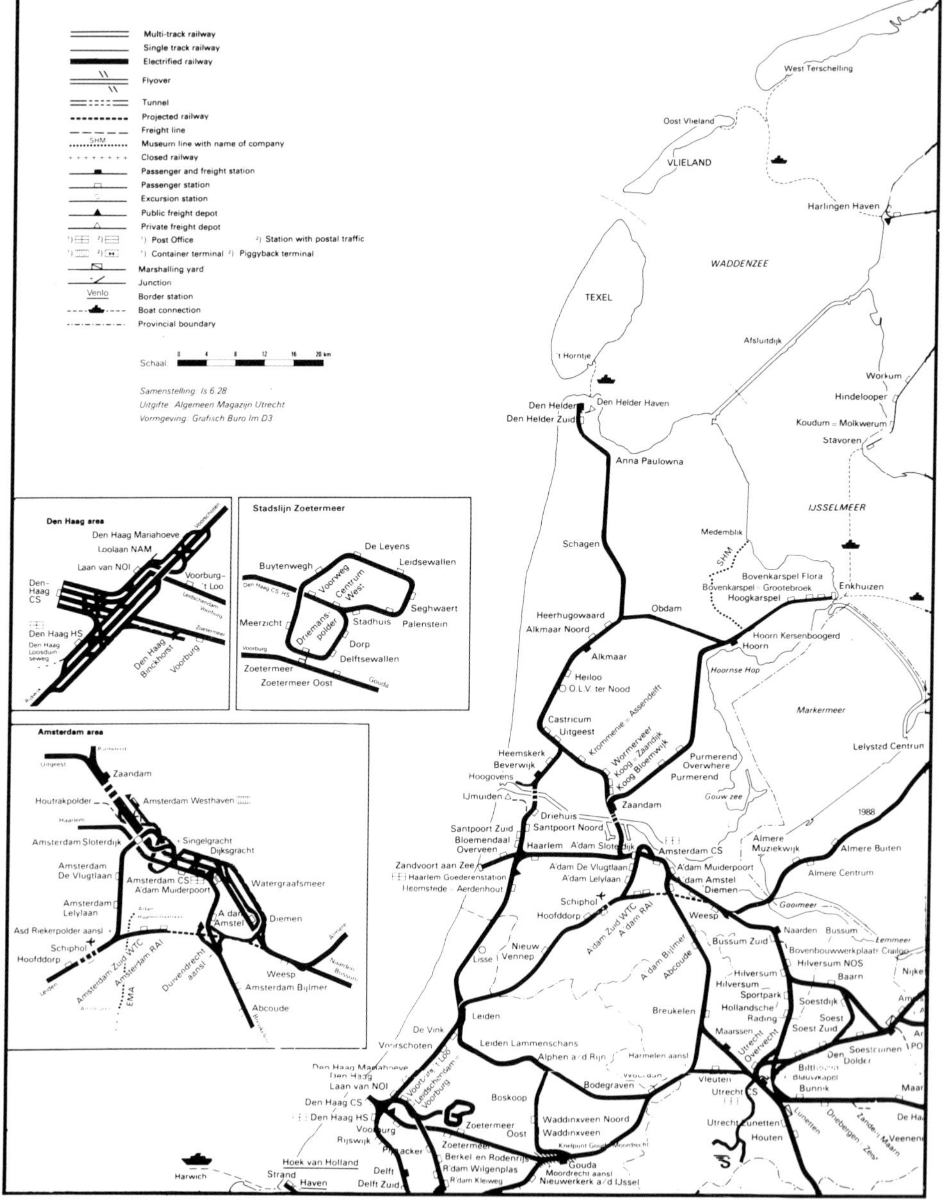

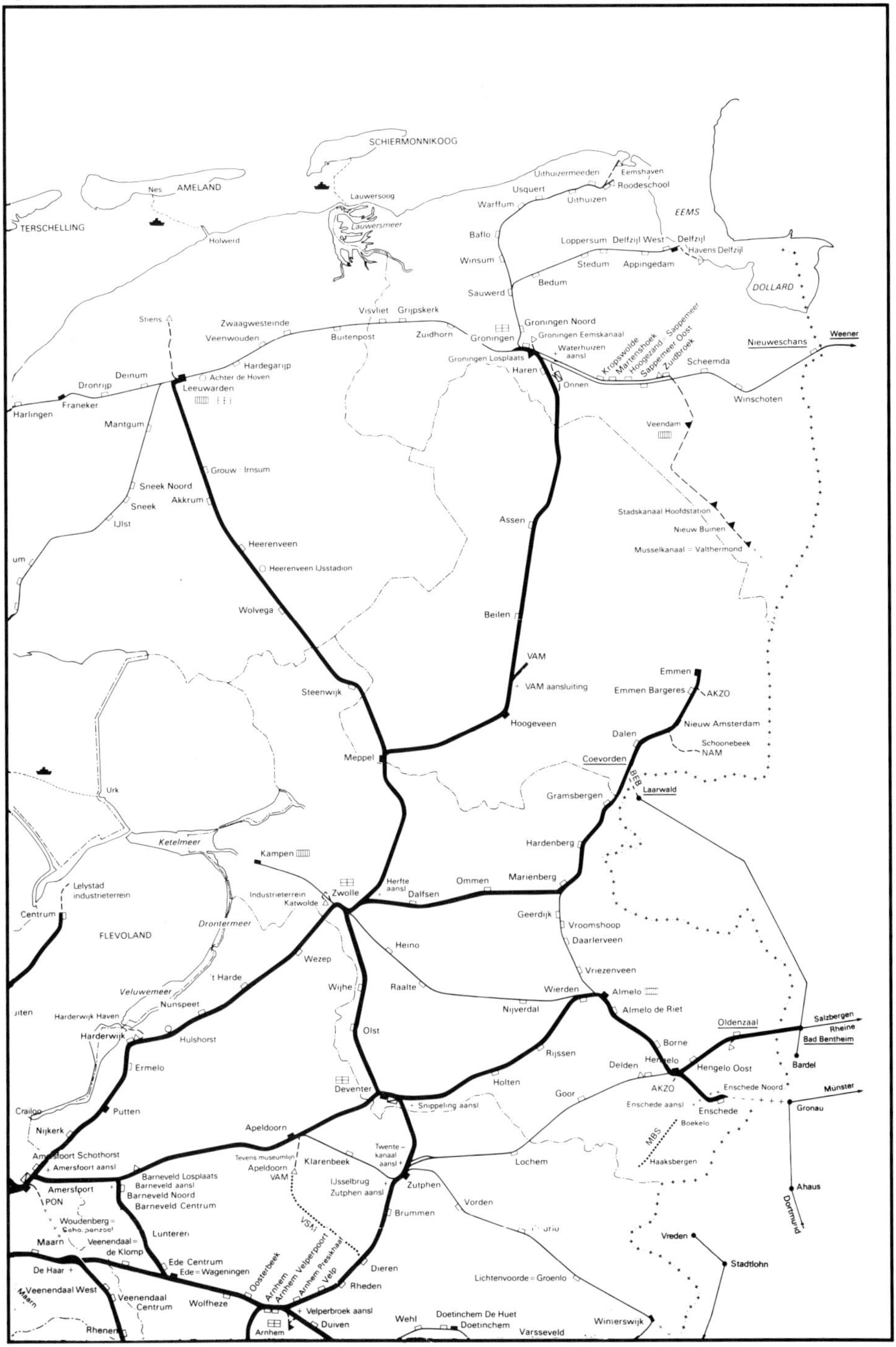

SCHIERMONNIKOOG
AMELAND
Nes
TERSCHELLING
Holwerd
Lauwersoog
Lauwersmeer
Uithuizermeeden
Eemshaven
Roodeschool
Usquert
Uithuizen
Warffum
EEMS
Baflo
Loppersum
Delfzijl West
Delfzijl
Havens Delfzijl
Winsum
Stedum
Appingedam
Bedum
DOLLARD
Sauwerd
Visvliet
Grijpskerk
Stiens
Zwaagwesteinde
Groningen Noord
Groningen Eemskanaal
Veenwouden
Buitenpost
Zuidhorn
Groningen
Waterhuizen aansl
Nieuweschans
Weener
Groningen Losplaats
Kropswolde
Martenshoek
Hoogezand Sappemeer
Sappemeer Oost
Zuidbroek
Scheemda
Hardegarijp
Haren
Deinum
Achter de Hoven
Onnen
Dronrijp
Leeuwarden
Franeker
Winschoten
Harlingen
Mantgum
Veendam
Grouw Irnsum
Sneek Noord
Akkrum
Sneek
Assen
Stadskanaal Hoofdstation
IJlst
Nieuw Buinen
Heerenveen
Musselkanaal = Valthermond
Heerenveen IJsstadion
Wolvega
Beilen
VAM
Emmen
VAM aansluiting
Emmen Bargeres
AKZO
Steenwijk
Hoogeveen
Nieuw Amsterdam
Dalen
Schoonebeek NAM
Meppel
Coevorden
BEB
Laarwald
Gramsbergen
Urk
Ketelmeer
Hardenberg
Kampen
Lelystad industrieterrein
Herfte aansl
Ommen
Marienberg
Industrieterrein Katwolde
Zwolle
Dalfsen
Centrum
Geerdijk
Vroomshoop
FLEVOLAND
Drontermeer
Daarlerveen
Heino
Wezep
't Harde
Vriezenveen
Wijhe
Raalte
Veluwemeer
Wierden
Almelo
Nunspeet
Nijverdal
Almelo de Riet
Harderwijk Haven
Oldenzaal
Salzbergen
Rheine
Harderwijk
Hulshorst
Olst
Bad Bentheim
Borne
Rijssen
Hengelo
Ermelo
Delden
Hengelo Oost
Bardel
Holten
AKZO
Enschede Noord
Deventer
Goor
Münster
Putten
Snippeling aansl
Enschede aansl
Enschede
Gronau
Boekelo
Nijkerk
Apeldoorn
MBS
Twente-kanaal aansl
Amersfoort Schothorst
Tevens museumlijn Apeldoorn VAM
Klarenbeek
Lochem
Haaksbergen
Amersfoort aansl
Barneveld Losplaats
Barneveld aansl
IJsselbrug Zutphen aansl
Zutphen
Ahaus
Amersfoort
Barneveld Noord
Vorden
PON
Barneveld Centrum
Brummen
Dortmund
Woudenberg = Scherpenzeel
VSM
Lunteren
Vreden
Maarn
Veenendaal = de Klomp
Ede Centrum
Stadtlohn
Ede = Wageningen
Oosterbeek
Arnhem Velperpoort
Arnhem Presikhaaf
Dieren
De Haar
Velp
Rheden
Lichtenvoorde = Groenlo
Veenendaal West
Veenendaal Centrum
Wolfheze
Arnhem
Velperbroek aansl
Wehl
Doetinchem De Huet
Winterswijk
Rhenen
Duiven
Doetinchem
Varsseveld

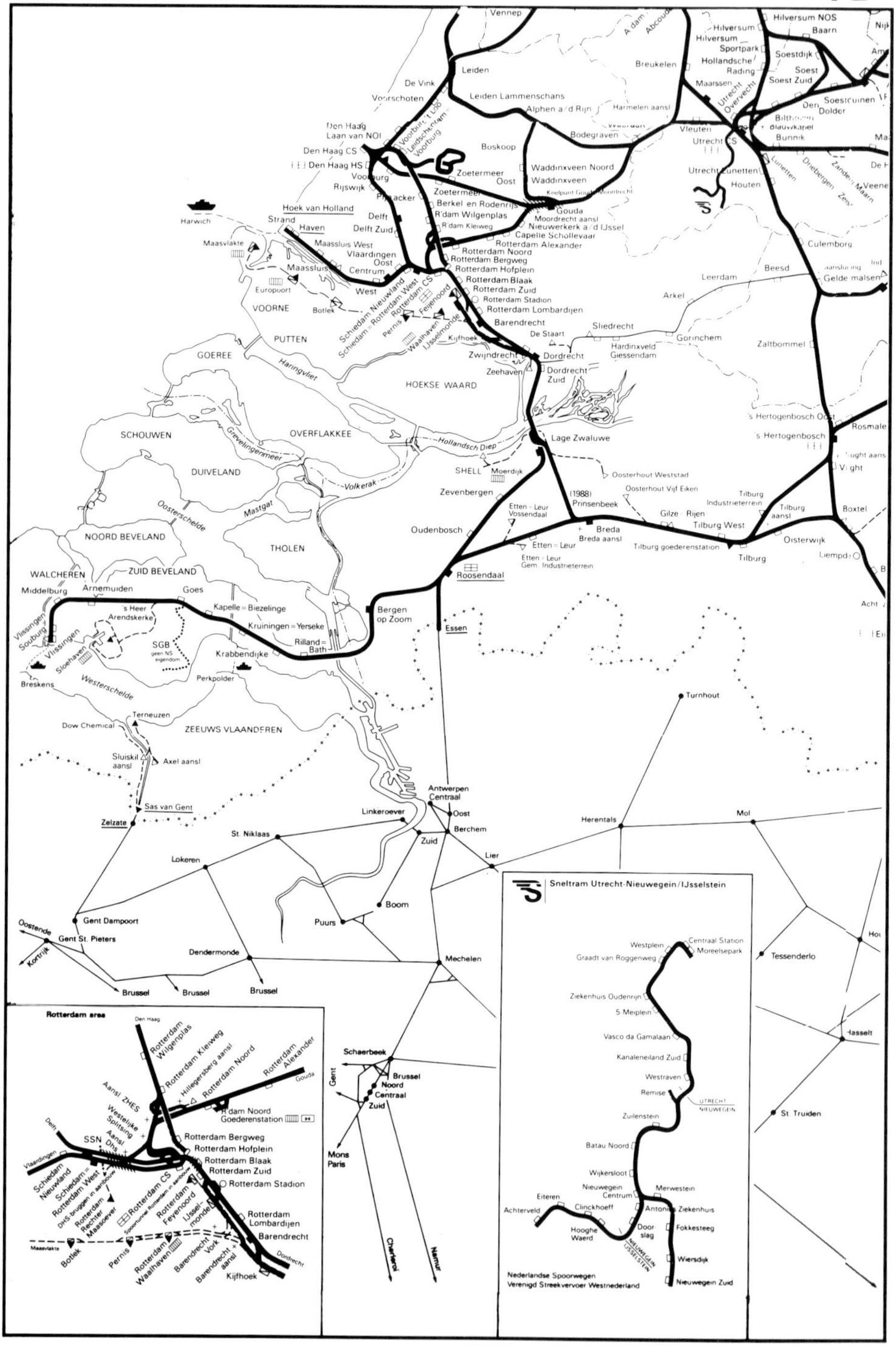
Rotterdam area
Sneltram Utrecht-Nieuwegein/IJsselstein
Nederlandse Spoorwegen
Verenigd Streekvervoer Westnederland
Leiden
Den Haag CS
Den Haag HS
Hoek van Holland
Gouda
Rotterdam CS
Rotterdam Blaak
Rotterdam Zuid
Dordrecht
Lage Zwaluwe
Breda
Roosendaal
Bergen op Zoom
Vlissingen
Middelburg
Goes
Tilburg
Utrecht CS
Hilversum
Amersfoort
's Hertogenbosch
Essen
Antwerpen Centraal
Mechelen
Brussel
Gent St. Pieters
Oostende
Kortrijk
Turnhout
Herentals
Mol
Hasselt
St. Truiden
Charleroi
Namur
Mons
Paris
SCHOUWEN
DUIVELAND
OVERFLAKKEE
GOEREE
VOORNE
PUTTEN
HOEKSE WAARD
NOORD BEVELAND
ZUID BEVELAND
WALCHEREN
THOLEN
ZEEUWS VLAANDEREN
Westerschelde
Oosterschelde
Haringvliet
Hollandsch Diep
Volkerak
Harwich

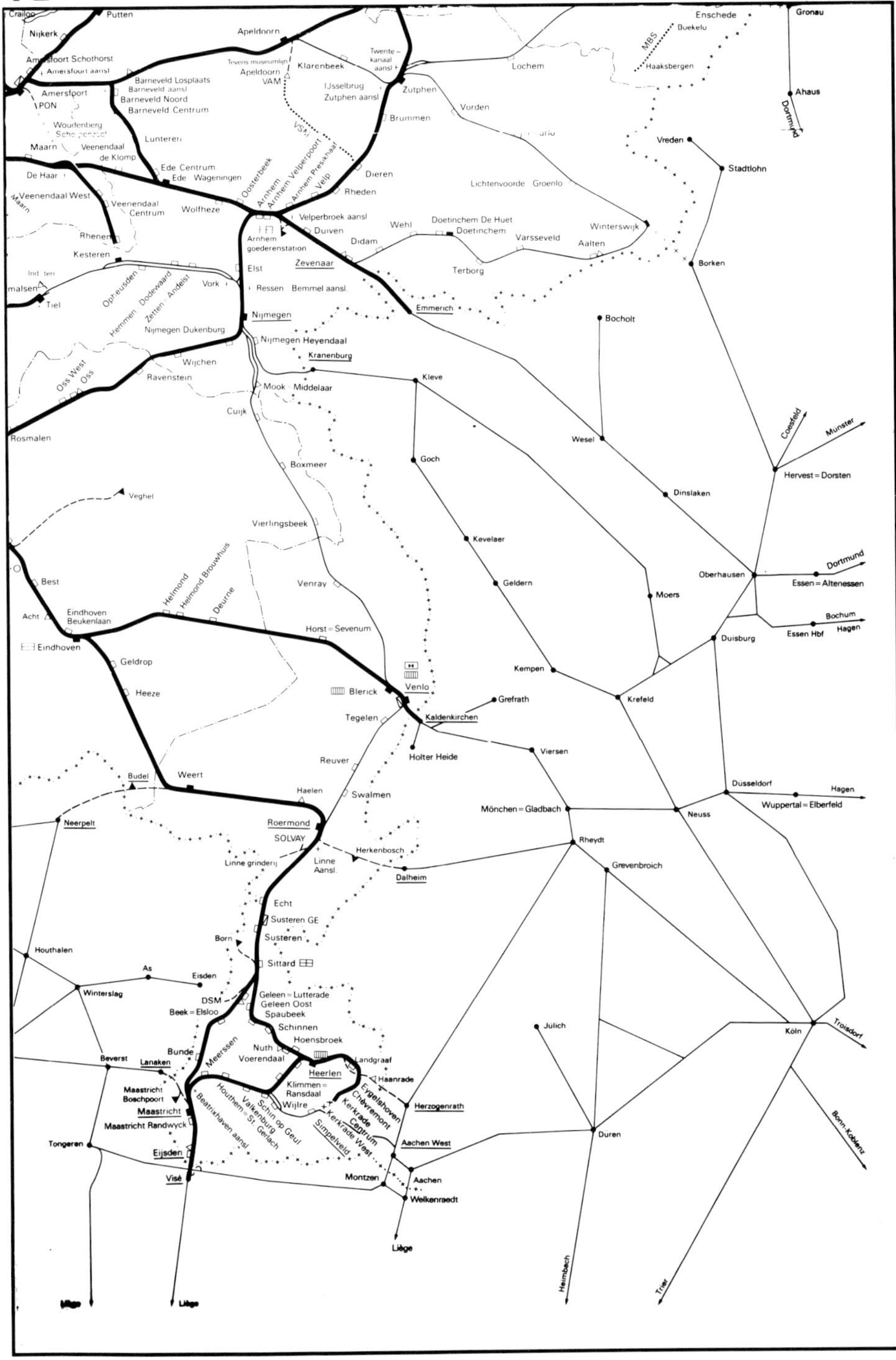

Putten
Nijkerk
Amersfoort Schothorst
Amersfoort aansl
Amersfoort
PON
Woudenberg
Maarn
Veenendaal de Klomp
De Haar
Veenendaal West
Veenendaal Centrum
Rhenen
Kesteren
Tiel
Apeldoorn
Klarenbeek
Tevens museumlijn
Apeldoorn VAM
Barneveld Losplaats
Barneveld aansl
Barneveld Noord
Barneveld Centrum
Lunteren
Ede Centrum
Ede Wageningen
Wolfheze
Oosterbeek
Arnhem
Arnhem Velperpoort
Arnhem Presikhaaf
Velp
Rheden
Dieren
Brummen
IJsselbrug
Zutphen aansl
Zutphen
Twente-kanaal aansl
Lochem
Vorden
Enschede
Boekelo
MBS
Haaksbergen
Gronau
Ahaus
Dortmund
Vreden
Stadtlohn
Lichtenvoorde Groenlo
Winterswijk
Aalten
Varsseveld
Terborg
Doetinchem De Huet
Doetinchem
Wehl
Didam
Duiven
Velperbroek aansl
Arnhem goederenstation
Zevenaar
Elst
Ressen Bemmel aansl.
Emmerich
Borken
Opheusden
Hemmen-Dodewaard
Zetten-Andelst
Vork
Nijmegen
Nijmegen Dukenburg
Nijmegen Heyendaal
Kranenburg
Kleve
Wijchen
Ravenstein
Oss West
Oss
Rosmalen
Mook Middelaar
Cuijk
Boxmeer
Bocholt
Wesel
Coesfeld
Munster
Hervest = Dorsten
Dinslaken
Goch
Veghel
Vierlingsbeek
Kevelaer
Geldern
Oberhausen
Dortmund
Essen = Altenessen
Moers
Bochum
Essen Hbf
Hagen
Duisburg
Best
Acht
Eindhoven Beukenlaan
Eindhoven
Helmond
Helmond Brouwhuis
Deurne
Venray
Horst = Sevenum
Geldrop
Heeze
Blerick
Venlo
Tegelen
Kaldenkirchen
Grefrath
Kempen
Krefeld
Holter Heide
Viersen
Reuver
Budel
Weert
Haelen
Swalmen
Neerpelt
Roermond
SOLVAY
Linne grinderij
Linne Aansl.
Herkenbosch
Dalheim
Mönchen = Gladbach
Neuss
Dusseldorf
Wuppertal = Elberfeld
Rheydt
Grevenbroich
Echt
Susteren GE
Susteren
Born
Sittard
Houthalen
As
Eisden
Winterslag
DSM
Beek = Elsloo
Geleen = Lutterade
Geleen Oost
Spaubeek
Schinnen
Hoensbroek
Nuth
Voerendaal
Landgraaf
Heerlen
Haanrade
Klimmen = Ransdaal
Wijlre
Bunde
Meerssen
Beverst
Lanaken
Maastricht Boschpoort
Maastricht
Maastricht Randwyck
Houthem = St. Gerlach
Valkenburg
Schin op Geul
Beatrixhaven aansl
Eygelshoven
Chèvremont
Kerkrade Centrum
Kerkrade West
Simpelveld
Herzogenrath
Aachen West
Aachen
Tongeren
Eijsden
Visé
Montzen
Welkenraedt
Liège
Julich
Duren
Köln
Troisdorf
Bonn-Koblenz
Heimbach
Trier

DIESEL SHUNTING LOCOMOTIVES

200 CLASS B

These small locomotives known as "Siks" (goats) are to be found on light duties in freight yards and engineers locations throughout the NS system. The light overall weight and robust construction have contributed to the longevity of the class. Formerly painted dark green with cast number plates.

Built: 1934–51.
Builder: Werkspoor (281–06 built by NS Zwolle Works).
Engine: Stork R153 of 63 kW.
Transmission: Electric. Two Heemaf TM6 or Smit GT 322/7 axle-hung.
Train Heating: None.
Maximum Tractive Effort: 39 kN.
Driving Wheel Dia.: 1000 mm.
Weight in Full Working Order: 21 (23 c) tonnes.
Length over Buffers: 7.22 m.
Max. Speed: 60 km/h.

c Fitted with telescopic crane.
t Fitted with radio telephone.

203	228	249	271	297	315	332	349
204	229	250 c	274 c	298	316	334	350
209	230	252 c	276 c	299	318	335	351
210	231	253	278	300	319	336	352
211	232	254	279	301	320	337	353
213	234	255	281	302	321	338	355 c
214	235	256	283	303	322	339	357
215	238	257	284 c	305	323	340	358
217	241	259	285	306	324	341	359
218	242 c	260	286	307	325	342	360 c
219	243	262	288	308	326	343	361 c
222	244	263	289	309 c	327	344	362 c
223	245	264	290	310	328	345	363
225	246 c	265 c	291	312	329	346	368 c
226	247	267	292	313	330	347	369
227 c	248 c	270	296	314	331	348	

500/600 CLASSES C

Instantly recognisable to the British eye, based on the BR classes 11 (500 class) and 08 (600 class) shunters of BR, these locomotives perform similar duties, considerable numbers being found in the larger yards and stations of the NS system. The 500 class have no train brake and are due for early withdrawal. 28 more 600s are being considered for fitting with remote control.

Built: 1950–57.
Builder: English Electric (Dick Kerr Works, Preston).
Engine: EE 6K of 260 kW (500 class) 6KT of 294 kW (600 class).
Transmission: Electric. Two EE 506 axle-hung traction motors.
Train Heating: None.
Maximum Tractive Effort: 143 kN.
Driving Wheel Dia.: 1230 mm.
Weight in Full Working Order: 47 tonnes.
Length over Buffers: 9.07 m.
Max. Speed: 30 km/h.

r Fitted with remote control.
t Fitted with radio telephone.

512	607	616	625	634	642	650 t	658
525	608	617	626 r	635	643 t	651 t	659
545	609 r	618	627	636	644	652	660 t
601	610	619	628	637	645 t	653 t	661 t
602	611	620	629	638 t	646	654	662
603	612	621	630	639	647 t	655 t	663
604	613	622	631	640	648 t	656	664 t
605	614	623	632	641	649 t	657	665
606	615	624	633 t				

ELECTRIC LOCOMOTIVES

1100 CLASS Bo–Bo

These locomotives are based on the SNCF BB 8100 class dating from 1949, modified with spring-borne traction motors. Alterations by the NS have included the fitting of roller bearings and larger sleeve-type buffers as carried by the 1600 class. From 1978 the most striking modification has been the addition of the new Alsthom-type nose ends, 1129 being the first so fitted and 1124 the last. 1101–50 were originally liveried in light blue, but this gave way in 1954 to dark blue with polished raised metal bands. Since 1971, the current NS yellow and grey livery has been applied to all locos. The locos work mixed-traffic duties all over the NS.

Built: 1950–56. (Rebuilt by NS 1978–82).
Builder-Mech. Parts: Alsthom.
Builder-Elec. Parts: Alsthom.
Traction Motors: 4 x Alsthom TA628A frame-mounted.
One Hour Rating: 2030 kW. **Weight**: 83 tonnes.
Maximum Tractive Effort: 152 kN. **Length over Buffers**: 14.110 m.
Driving Wheel Dia.: 1250 mm. **Max. Speed**: 135 km/h.

1101	1108	1114	1123	1134	1142	1148	1154
1102	1109	1115	1124	1135	1143	1149	1155
1103	1110	1116	1127	1136	1144	1150	1157
1104	1111	1117	1129	1137	1145	1151	1158
1105	1112	1119	1130	1138	1146	1152	1159
1106	1113	1122	1132	1139	1147	1153	1160
1107							

1200 CLASS Co–Co

The striking 1200 class, originally built for freight haulage were constructed as 'kit-form' locomotives with the bogies being supplied by Baldwin and electrical components by Westinghouse, the clasic American styling clearly showing their design origin. Until refurbishment by the NS, the mixed parentage was displayed on elaborate worksplates carried on the nose ends. These locos, extremely popular with their NS crews, originally carried the light blue (1215 mahogany!) and dark blue liveries but all are now in the standard colours of yellow and grey.

Built: 1951–53.
Builder-Mech. Parts: Werkspoor.
Builder-Elec. Parts: Heemaf.
Traction Motors: 4 x Heemaf TM94 axle-hung.
One Hour Rating: 2360 kW. **Weight**: 108 tonnes.
Maximum Tractive Effort: 194 kN. **Length over Buffers**: 18.085 m.
Driving Wheel Dia.: 1100 mm. **Max. Speed**: 135 km/h.

1201	1205	1208	1211	1214	1217	1220	1223
1202	1206	1209	1212	1215	1218	1221	1224
1203	1207	1210	1213	1216	1219	1222	1225
1204							

1300 CLASS Co–Co

The impressive-looking 1300 class are based on the SNCF CC-7100 class which includes in it's ranks CC 7107, the co-world speed record holder (265 km/h attained in 1955). The NS 1300 duties are much less glamorous, being generally utilised for heavy freight work with limited passenger diagrams. One example carries 'elbow' style pantographs instead of the standard 'box' type. The class has recently undergone a major refurbishment programme, the external appearance being more in keeping with the class 1600 livery but with an extremely large NS logo on the body side and an EMU-style lamp assembly on the front end.

Built: 1952–56.
Builder-Mech. Parts: Alsthom.
Builder-Elec. Parts: Alsthom.
Traction Motors: 6 x Alsthom TA 628A frame mounted.

One Hour Rating: 3045 kW.
Maximum Tractive Effort: 226 kN.
Driving Wheel Dia.: 1250 mm.
Weight: 111 tonnes.
Length over Buffers: 18.952 m.
Max. Speed: 135 km/h.

1301
1302 WOERDEN
1304
1305 ALPHEN AAN DER RIJN
1306
1307
1308
1309
1310 BUSSUM
1311
1312 ZOETERMEER
1313
1314 HOORN
1315 TIEL
1316 GELDERMALSEN

1600 CLASS B–B

Named after towns on the NS electrified lines, the chopper-controlled 1600 class represents a major investment for the NS. The bright yellow locomotives are based on the SNCF BB 7200 class designed for a maximum speed of 200 km/h (limited at present to 160 km/h). The locos work mixed traffic duties over all principal routes.

Built: 1981–83.
Builder-Mech. Parts: Alsthom.
Builder-Elec. Parts: Alsthom.
Traction Motors: 2 x Alsthom TAB 674 C4 frame mounted.
One Hour Rating: 4400 kW.
Maximum Tractive Effort: 294 kN.
Driving Wheel Dia.: 1250 mm.
Weight: 83 tonnes.
Length over Buffers: 17.48 m.
Max. Speed: 160 km/h.

All push-pull fitted.

1601 AMSTERDAM
1602 Schiphol
1603 ZUTPHEN
1604 DORDRECHT
1605 BREDA
1606 HARDERWIJK
1607 VLISSINGEN
1608 S'HERTOGENBOSCH
1609 HOOFDDORP
1610 HENGELO
1611 VENLO
1612 GOES
1613 ROERMOND
1614 SCHIEDAM
1615 ZANDVOORT
1616 OLDENZAAL
1617 ASSEN
1618 ALMELO
1619 MAASTRICHT
1620 ARNHEM
1621 DEVENTER
1622 HAARLEM
1623 HILVERSUM
1624 ALKMAAR
1625 SITTARD
1626 MEPPEL
1627 GOUDA
1628 APELDOORN
1629 EDE
1630 ZWOLLE
1631 VOORBURG
1632 NIJMEGEN
1633 BERGEN OP ZOOM
1634 LELYSTAD
1635 ENSCHEDE
1636 HEEREVEEN
1637 AMERSFOORT
1638 GRONINGEN
1639 LEIDEN
1640 STEENWIJK
1641 ALMERE
1642 WEERT
1643 HEERLEN
1644 ROOSENDAAL
1645 MIDDELBURG
1646 LEEUWARDEN
1647 DELFT
1648 VALKENBURG
1649 OSS
1650 DEN HAAG
1651 TILBURG
1652 UTRECHT
1653 DEN HELDER
1654 GELEEN
1655 EINDHOVEN
1656 HOOGEVEEN
1657 ROTTERDAM
1658 ZAANDAM

1700 CLASS B–B

These locos on order are expected to be similar to class 1600.

1701
1702
1703
1704
1705
1706
1707
1708
1709
1710
1711
1712
1713
1714
1715
1716
1717
1718
1719
1720
1721
1722
1723
1724

1725	1727	1729	1731	1733	1735	1737	1738
1726	1728	1730	1732	1734	1736		

MAIN-LINE DIESEL LOCOMOTIVES

2200 CLASS Bo–Bo

These locos which have a cab at one end can be found either double-heading or singly all over the NS system on freight work. They also work into West Germany. Radio-fitted examples work in Kijfhoek Yard near Rotterdam, Waalhaven Zuid, Botkek and Dordrecht.

Built: 1955–58.
Builder: Allen (2201–2300), Schneider (2301–2350).
Engine: Stork Schneider Superior 40C-Lx-8 of 670 kW at 1100 rpm.
Transmission: Electric. 4 Heemaf TM98 traction motors.
Train Heating: None.
Maximum Tractive Effort: 181 kN.
Driving Wheel Dia.: 950 mm.
Weight in Full Working Order: 72 tonnes.
Length over Buffers: 14.100 m.
Max. Speed: 100 km/h.

r Fitted with remote control.
t Fitted with radio telephone.

2201	2217	2236	2255	2272 t	2291	2306	2328
2202	2219	2238	2256	2273 t	2292	2307	2329
2203	2220	2239	2257	2274 t	2293	2308	2330
2204	2221	2241	2258	2275 **R** t	2294	2309	2332
2205	2222	2243 r	2259	2276 t	2295	2310	2333
2207	2223	2244	2260 r	2277 t	2296	2311	2334
2208	2224	2245	2261 r	2278 t	2297	2317	2335
2209	2225	2246	2263 r	2280	2298	2318	2336
2210	2226	2247	2264 r	2282 t	2299	2319	2338
2211	2228	2249	2266 r	2283 t	2300	2320	2339
2212	2229	2250	2267 r	2284	2301 t	2322	2340
2213	2230	2251	2268 r	2285	2302 t	2323	2342
2214	2233	2252	2269 r	2286	2303 t	2324	2344
2215	2234	2253	2270 r	2289	2304 t	2326	2346
2216	2235	2254	2271 rt	2290	2305	2327	2348

2400 CLASS Bo–Bo

These locos with an outward resemblance to the former class 15 (BTH type 1) locos of BR see use on freight duties over the whole of the NS system. The numbers in service have been depleted recently due to the diminishing wagon-load traffic. The class will eventually be replaced by the 6400 class.

Built: 1954–56.
Builder: Alsthom.
Engine: SACM V12 SHR of 625 kW.
Transmission: Electric. 4 x Alsthom TA637 axle-hung traction motors.
Train Heating: None.
Maximum Tractive Effort: 161 kN.
Driving Wheel Dia.: 1000 mm.
Weight in Full Working Order: 60 tonnes.
Length over Buffers: 12.520 m.
Max. Speed: 80 km/h.

2401	2413	2429 h	2444	2453	2467	2482	2509
2403	2414	2430	2446	2454	2470	2501 **R**	2510
2404	2416	2434	2447	2456	2471	2502	2515
2405	2419	2435	2448	2457	2477	2504	2518
2406	2423	2439	2449	2458	2478	2506	2522
2411	2424	2441	2450	2463	2481	2508	2526
2412	2425 h	2443	2452	2465			

2530 CLASS Bo–Bo

The sole representative of its type, this locomotive is a 2400 class with a different body design. Details as for 2400 class except: Length over Buffers: 13.30 m.

2530

▲**Class 25 No. 2508 stands at Zutphen on 30/07/88.** *Alex Dasi-Sutton*

▼**The unique No. 2530 at Utrecht on 06/04/86. This loco is at present allocated to weedkilling train duties, and may be withdrawn before the end of 1989.** *Alex Dasi-Sutton*

6400 CLASS Bo–Bo

New thyristor-controlled locomotives for freight and shunting use.

Built: 1988 onwards.
Builder: MaK.
Engine: MaK of 1180 kW at 1800 rpm.
Transmission: 4 x three phase BBC traction motors.
Train Heating: None.
Maximum Tractive Effort: kN.
Driving Wheel Dia.: mm.
Weight in Full Working Order: 80 tonnes.
Length over Buffers: 14.40 m.
Max. Speed: 120 km/h.

6401 Mijndert	6416	6431	6446
6402 Marinus	6417	6432	6447
6403 Gijs	6418	6433	6448
6404 Johan	6419	6434	6449
6405 Jan	6420	6435	6450
6406 Tonnie	6421	6436	6451
6407 Henk	6422	6437	6452
6408	6423	6438	6453
6409	6424	6439	6454
6410	6425	6440	6455
6411	6426	6441	6456
6412	6427	6442	6457
6413	6428	6443	6458
6414	6429	6444	6459
6415	6430	6445	6460

NS WITHDRAWN LOCOMOTIVES

The following are the last known locations of withdrawn locomotives which have not yet been scrapped:

523 Roosendaal	543 Roosendaal	2440 Blerick	2495 Roosendaal
524 Roosendaal	2407 Amersfoort	2459 Blerick	2502 Roosendaal
528 Roosendaal	2409 Roosendaal	2462 Zwolle	2513 Roosendaal
531 Roosendaal	2410 Roosendaal	2466 Roosendaal	2516 Blerick
534 Roosendaal	2415 Tilburg	2481 Roosendaal	2518 Blerick
538 Roosendaal	2431 Roosendaal	2491 Roosendaal	2529 Roosendaal
542 Roosendaal	2432 Venlo	2493 Roosendaal	

▼ Plan U 'DE III' DEMU No. 134 stands in the sidings at Zutphen waiting to form the 13.13 service to Enschede on 30/07/88. *David Brown*

DIESEL MULTIPLE UNITS

VIP CAR ('THE CAMEL') SINGLE UNIT

The ex-directors saloon nicknamed 'The Camel' because of its two humps is available for party hire and may be seen anywhere on the NS. It is also permitted to work in Belgium, Luxembourg and West Germany. The yellow livery is enhanced by a blue stripe and the raised driving positions and lack of auto-coupling make this a very distinctive vehicle.

Wheel Arrangement: Bo–Bo.
Built: 1954.
Builder: Allan.
Engine: AEC A220 of 134 kW.
Transmission: Electric.
Seats: 24F 2L.
Weight: 58 tonnes.
Length over couplings: 25.55 m.
Max. Speed: 125 km/h.

20

CLASS DEIII (PLAN U) 3-CAR UNITS

In many respects similar to the 1962 stock EMUs, these three-car 'dog head' EMUs work over the following lines: Dordrecht–Geldermalsen, Roermond–Nijmegen, Zwolle–Enschede/Emmen, and Arnhem–Winterswijk. Automatic doors. PA.

BD+B+AB (DMBSO–TSOL–DTCsoL).

Built: 1960–63.
Builder: Werkspoor.
Wheel Arrangement: Bo–Bo+2–2+2–2.
Engine: SACM MGO-12-BSHR of 735 kW.
Transmission: Electric. Smit GT 38/224 or Metropolitan Vickers MV 139.
Seats: 40S + 88S 2L + 24F 40S 2L.
Weight: 66 + 35 + 35 tonnes.
Length over couplings: 25.17 + 24.09 + 25.17 m.
Max. Speed: 125 km/h.

111	117	123	128	133	138	143	148
112	118	124	129	134	139	144	149
113	119	125	130	135	140	145	150
114	120	126	131	136	141	146	151
115	121	127	132	137	142	147	152
116	122						

CLASS DEII (PLAN X-v) 2-CAR UNITS

These articulated units (ex 61–106 series) were originally liveried in light blue and were nicknamed "blue angels". They have been extensively rebuilt and modernised and have large pods on the roof for pressure ventilation equipment. Duties include Zwolle–Kampen, Arnhem–Nijmegen–Teil, Apeldoorn–Winterswijk and Maastricht–Arnhem–Tiel.

AB+B (DMCOL–DMSOL).

Built: 1953–54. Rebuilt 1975/77–82.
Builder: Allan. Rebuilt NS Haarlem.
Wheel Arrangement: Bo–2–Bo.
Engine: Cummins NT895R2 of 180 kW (one per car).
Transmission: Electric.
Seats: 16F 56S 1L + 76S 1L. **Weight:** 45 + 45 tonnes.
Length over couplings: 22.70 + 22.70 m. **Max. Speed:** 120 km/h.

162 (92)	167 (61)	171 (98/100)	175 (68/64)	179 (72/73)	183 (67)
163 (83)	168 (91)	172 (66/86)	176 (64/62)	180 (73/66)	184 (63/65)
164 (106/69)	169 (94)	173 (100/70)	177 (62/82)	181 (99)	185 (102)
165 (103)	170 (69/98)	174 (70/68)	178 (82/72)	182 (75)	186 (77/87)
166 (97)					

3100/3200 CLASSES:

These units represented a change in policy for the NS. The power equipment was based on that used to modernise the DEII units, i.e. a Cummins engine, but Voith hydraulic transmission was used. A new body style was adopted with flat ends with one-piece windscreens. The units work on branch lines out of Groningen, and carry the logo 'Wadloper'. Literally translated this means 'mud-flat runner' and refers to the terrain of the area.

3100 CLASS — SINGLE UNITS

B (DMSOL).

Built: 1983.
Builder: Duewag.
Wheel Arrangement: 2–B.
Engine: Cummins NT855R4 of 210 kW.
Transmission: Hydraulic. Voith T211r.
Seats: 56S 1L.
Length over couplings: 22.31 m.
Weight: 36 tonnes.
Max. Speed: 100 km/h.

Disc brakes. Magnetic track brakes.

3101	3104	3107	3110	3112	3114	3116	3118
3102	3105	3108	3111	3113	3115	3117	3119
3103	3106	3109					

3200 CLASS — 2-CAR UNITS

B+B (DMSOL–DMSO).

Built: 1981–82.
Builder: Duewag.
Wheel Arrangement: 2–B+B–2.
Engine: Cummins NT855R4 of 210 kW (one per car).
Transmission: Hydraulic. Voith T211r.
Seats: 68S 1L + 72S.
Length over couplings: 21.72 + 21.72 m.
Weight: 35 + 35 tonnes.
Max. Speed: 100 km/h.

Disc brakes. Magnetic track brakes.

3201	3205	3209	3213	3217	3221	3225	3228
3202	3206	3210	3214	3218	3222	3226	3229
3203	3207	3211	3215	3219	3223	3227	3230
3204	3208	3212	3216	3220	3224		

▼Plan X-v rebuilt 'DE II' 2-car DEMU No. 173 arrives at Almelo with the 12.28 service from Zwolle on 25/08/87.
David Brown

ELECTRIC MULTIPLE UNITS

All NS EMUs are gangwayed within the unit only and have tread brakes except where stated otherwise.
The various builds of NS EMUs can easily be recognised by their front-end design as follows:
1946 stock ('Mat 46') had a round front and a characteristic green livery. All 1946 stock is now withdrawn from capital stock.
1954 stock ('Mat 54') has a large 'dog head' and was designed for Inter-City services.
1962 stock ('Mat 62') has a short 'dog head' with the outer cab windows pointed. It is used on stopping services.
Sprinter stock has a slightly sloping front end.
IC3 stock has gangwayed ends with a roof cab.

EMUs are often referred to by their formation code as follows:
EL EMU, D includes guard's/luggage area (dienst), P Post area. 2, 3 or 4 No. of cars.
Example: ELD-2 is a two car EMU with guard's/luggage accommodation.

PLAN F,G, M & Q — 2-CAR UNITS

1954 stock. These sets are used on local routes such as Utrecht CS–Teil/Rhenen and Amsterdam CS–Arnhem and to strenthen 711 series workings (q.v.). Automatic doors. PA.

ABD+B (DMBCOL–DMSsoL).

Built: 1956–58.
Builder-Mech. Parts: Allan (Werkspoor*).
Builder-Elec. Parts: Smit.
Wheel Arrangement: Bo–2+2–Bo.
Traction Motors: 4 x 170 kW.
Accommodation: 24F 24S 1L + 72S 1L.
Weight: 56 + 54 (54 + 50*) tonnes.
Length over couplings: 25.56 + 25.56 m.
Max. Speed: 140 km/h.

321–334 are plan F, 335–350 are plan G, 351–365 are plan M and 371–393 are plan Q.

321	331	339	347	356	364	377 *	386 *
322	332	340	348	357	365	378 *	387 *
324	333	341	350	358	371 *	379 *	388 *
325	334	342	351	359	372 *	381 *	389 *
326	335	343	352	360	373 *	382 *	390 *
327	336	344	353	361	374 *	383 *	391 *
328	337	345	354	362	375 *	384 *	392 *
329	338	346	355	363	376 *	385 *	393 *
330							

PLAN V1, V2 & V3 — 2-CAR UNITS

1962 stock. These sets perform similar duties to the 321 series on local stopping trains on routes such as Amsterdam CS–Arnhem, Rotterdam CS–Nijmegen and Utrecht CS–Leiden.

ABD+B (DMBCOL–DMSsoL).

Built: 1966–68.
Builder-Mech. Parts: Werkspoor.
Builder-Elec. Parts: Smit.
Wheel Arrangement: 2–Bo+Bo–2.
Traction Motors: 4 x Heemaf of 145 kW.
Accommodation: 24F 24S 2L + 80S 1L.
Weight: 43 + 42 tonnes.
Length over couplings: 26.07 + 26.07 m.
Max. Speed: 140 km/h.

401–415 are plan V1, 416–430 are plan V2 and 431–438 are plan V3.

401	405	409	413	417	421	425	429
402	406	410	414	418	422	426	430
403	407	411	415	419	423	427	431
404	408	412	416	420	424	428	432

433	434	435	436	437	438

CLASS V4, V5 & V6 — 2-CAR UNITS

1962 stock. These sets are similar to 401–438 but have no luggage compartment. They work very short-distance traffic such as Amsterdam CS–Hoorn/Amersfoort and Utrecht CS–Leiden/ Baarn/ Amsterdam CS.

AB+B (DMCsoL–DMSsoL).

Built: 1969–70.
Builder-Mech. Parts: Werkspoor (Talbot*).
Builder-Elec. Parts: Smit.
Wheel Arrangement: 2–Bo+Bo–2.
Traction Motors: 4 x Heemaf 145 kW.
Accommodation: 24F 40S 1L + 78S 1L.
Weight: 43 + 42 tonnes.
Length over couplings: 26.07 + 26.07 m.
Max. Speed: 140 km/h.

441–461 are V4, 462–471 are V5 and 472–483 are V6.

441	447	453	459	464 *	469 *	474 *	479 *
442	448	454	460	465 *	470 *	475 *	480 *
443	449	455	461	466 *	471 *	476 *	481 *
444	450	456	462 *	467 *	472 *	477 *	482 *
445	451	457	463 *	468 *	473 *	478 *	483 *
446	452	458					

PLAN TT — 4-CAR UNIT

Prototype unit for 1962 stock. Automatic doors. PA.

B+AD+B+B (DTSsoL–MBFsoL–MSOL–DTSsoL).

Built: 1961.
Builder-Mech. Parts: Werkspoor.
Builder-Elec. Parts: Smit.
Wheel Arrangement: 2–2+Bo–Bo+Bo–Bo+2–2.
Traction Motors: 8 x Heemaf 150 kW.
Accommodation: 80S 1L + 41F 1L + 80S1L + 80S 1L.
Weight: 38 + 46 + 44 + 35 tonnes.
Length over couplings: 25.70 + 24.27 + 24.27 + 25.70 m.
Max. Speed: 140 km/h.

501

PLAN T — 4-CAR UNITS

These sets operate over a more limited area being normally confined to the Dordrecht–Den Haag–Amsterdam CS line and also the Den Haag–Venlo services which are not loco-hauled.

B+BD+AB+B (DTSsoL–MBSOL(K)–MCsoL–DTSoL).

Built: 1964–65.
Builder-Mech. Parts: Werkspoor.
Builder-Elec. Parts: Smit.
Wheel Arrangement: 2–2+Bo–Bo+Bo–Bo+2–2.
Traction Motors: 8 x Heemaf 150 kW.
Accommodation: 80S 1L + 24S 22 buffet 1L + 42F 24S 1L + 80S 1L.
Weight: 39 + 47 + 46 + 36 tonnes.
Length over couplings: 26.07 + 24.93 + 24.93 + 26.07 m.
Max. Speed: 140 km/h.

502	506	510	514	518	522	526	529
503	507	511	515	519	523	527	530
504	508	512	516	520	524	528	531
505	509	513	517	521	525		

▲Plan T 1962 stock 4-car EMU No. 507 slows to call at Barendrecht whilst forming the 09.10 Amsterdam CS–Dortrecht service on 25/07/88. *David Brown*

▼A pair of 1954 stock 'Hondekop' (dog-head) EMUs Nos. 1768 and 712 pass Den Haag Laan van N.O.I. with the 14.56 Amsterdam CS–Vlissingen service on 15/08/86. 1768 is in Inter-City livery, whilst 712 is in plain yellow. *David Brown*

PLAN F, G & P — 4-CAR UNITS

1954 stock. These units work indiscriminately with the refubished variety, principally on the Amsterdam–Vlissingen, Amsterdam–Arnhem, Zwolle–Roosendaal and Den Haag–Venlo services. They are often strengthened with 2-car series 321 units. An extended program of removal of blue asbestos from these units will lead to certain services reverting to loco-haulage (see earlier).

B+A+B+BD (DMSsoL–TFsoL–TSOL(K)–DMBSOL).

Built: 1956–62.
Builder-Mech. Parts: Werkspoor, Allan & Beijnes.
Builder-Elec. Parts: Smit.
Wheel Arrangement: Bo–Bo+2–2+2–2+Bo–Bo.
Traction Motors: 8 x Heemaf 170 kW.
Accommodation: 72S (70S*†) 2L + 48F 1L + 56S 22 buffet 2L + 40S
Weight: 59 + 48 + 50 + 56 (57 + 46 + 48 + 55†) tonnes.
Length over couplings: 25.56 + 24.07 + 24.07 + 25.56 m.
Max. Speed: 140 km/h.

711–741 were plan F, 742–757 plan G and 761–786 plan P. Many have been refurbished and renumbered in the 17xx series.

712	717	723	727	736	742 *	753 *	770 †
713	718	724	728	738	747 *	754 *	771 †
714	719	725	730	739	751 *	756 *	772 †
715	720	726	734	740	752 *	769 †	774 †
716	722						

PLAN V7 — 2-CAR UNITS

1962 stock. Similar to other plan V units, but with post compartment.

AB+BP (DMCsoL–DMPSOL).

Built: 1970–72.
Builder-Mech. Parts: Werkspoor.
Builder-Elec. Parts: Smit.
Wheel Arrangement: 2–Bo+Bo–2.
Traction Motors: 4 x Heemaf 145 kW.
Accommodation: 24F 40S 1L + 56S.
Weight: 43 + 42 tonnes.
Length over couplings: 26.07 + 26.07 m.
Max. Speed: 140 km/h.

801	806	811	816	821	826	831	836
802	807	812	817	822	827	832	837
803	808	813	818	823	828	833	838
804	809	814	819	824	829	834	839
805	810	815	820	825	830	835	840

PLAN V8–V13 — 2-CAR UNITS

1962 stock. Talbot version of plan V7.

AB+BP (DMCsoL–DMPSOL).

Built: 1972–76.
Builder-Mech. Parts: Talbot.
Builder-Elec. Parts: Smit.
Wheel Arrangement: 2–Bo+Bo–2.
Traction Motors: 4 x Heemaf 145 kW.
Accommodation: 24F 40S 1L + 56S.
Weight: 45 + 43 tonnes.
Length over couplings: 26.07 + 26.07 m.
Max. Speed: 140 km/h.

841	844	847	850	853	856	859	862
842	845	848	851	854	857	860	863
843	846	849	852	855	858	861	864

865	879	892	905	917	929	941	953
866	880	893	906	918	930	942	954
867	881	894	907	919	931	943	955
868	882	895	908	920	932	944	956
869	883	896	909	921	933	945	957
870	884	897	910	922	934	946	958
871	885	898	911	923	935	947	960
872	886	899	912	924	936	948	961
873	887	900	913	925	937	949	962
874	888	901	914	926	938	950	963
875	889	902	915	927	939	951	964
876	890	903	916	928	940	952	965
877	891	904					

PLAN F-v, G-v & P-v — 4-CAR UNITS

1954 stock. These units were rebuilt in 1975–78 with unidirectional seating in second class. Details as for 712–774 except:
Accommodation: 64S (62S*) + 48F 1L + 48S 22 buffet 2L + 48S 2L.

One thousand has been added to former numbers.

1711 I	1733 I	1741 I	1745 I	1749 I *	1757 I *	1763 I *	1766 I *
1721 I	1735 I	1743 I	1746 I	1750 I *	1761 I *	1764 I *	1767 I *
1729 I	1737 I	1744 I	1748 I *	1755 I *	1762 I *	1765 I *	1768 I *
1731 I							

PLAN P-v — 4-CAR UNITS

These are an earlier rebuild and also have unidirectional seating in the first class. They also have all-over blue doors.

Accommodation: 58S + 45F 1L + 48S 22 buffet 2L + 32S 2L.

1780 (780) I	1782 (782) I	1784 (784) I	1786 (786) I	1788 (775) I	1790 (778) I
1781 (781) I	1783 (783) I	1785 (785) I	1787 (773) I	1789 (777) I	1791 (779) I

'TREINSTEL 1970' — 4-CAR UNIT

1954 stock. Rebuilt 1970 as a prototype for the future. Details as for 761–786 except:

B+A+Ar+A (DMSsoL–FsoL–TFOLRB–DMFO). The buffet car seats are now declassified.

Accommodation: 58S + 48F 1L + 31F 22 buffet 2L + 45F.

1970 (776) I

PLAN Y0 (SPRINTER SGM0) — 2-CAR UNITS

The earliest batch of 'Sprinter' units from which BR's new DMUs take their name, these sets work on the Zoetermeer Stadslijn which is a circular route through the eastern suburbs of Den Haag. The rapid acceleration from a standing start is especially useful on this line which has stations at very close intervals.

AB+B (DMCO–DMSO).

Built: 1975–76.
Builder-Mech. Parts: Talbot.
Builder-Elec. Parts: Oerlikon.
Wheel Arrangement: Bo–Bo+Bo–Bo.
Traction Motors: 8 x Oerlikon 160 kW.
Accommodation: 32F 40S + 72S.
Weight: 52.5 + 52.5 tonnes.
Length over couplings: 26.10 + 26.10 m.
Max. Speed: 125 km/h.

Disc brakes. Non-gangwayed. Automatic doors. PA.

2001	2002	2003	2004	2005	2006	2007	2008

2009	2010	2011	2012	2013	2014	2015	

PLAN Y1 (SPRINTER SGM1) 2-CAR UNITS

Similar to plan Y0, but with a toilet and gangways. Similar workings.

AB+B (DMCO–DMSOL).

Built: 1975–76.
Builder-Mech. Parts: Talbot.
Builder-Elec. Parts: Oerlikon.
Wheel Arrangement: Bo–Bo+Bo–Bo.
Traction Motors: 8 x Oerlikon 160 kW.
Accommodation: 32F 40S + 72S 1L.
Weight: 54 + 53 tonnes.
Length over couplings: 26.10 + 26.10 m.
Max. Speed: 125 km/h.

Disc brakes. Automatic doors. PA.

2021	2023	2025	2027	2029	2031	2033	2035
2022	2024	2026	2028	2030	2032	2034	

PLAN Y2/Y3 (SPRINTER SGM 1/2) 3-CAR UNITS

These units are generally confined to workings in the west of the country around Rotterdam and Den Haag and work such turns as Den Haag CS–Leiden–Amsterdam CS, Den Haag CS–Rotterdam CS–Dordrecht–Roosendaal, Den Haag–Rotterdam Hofplein and Rotterdam CS–Hoek van Holland.

B+AB+B (DMSO–TCO–DMSO).

Built: 1978–80/83–84*.
Builder-Mech. Parts: Talbot.
Builder-Elec. Parts: Oerlikon.
Wheel Arrangement: Bo–Bo+2–2+Bo–Bo.
Traction Motors: 8 x Oerlikon 160 kW.
Accommodation: 72S + 40F 40S + 72S 1L.
Weight: 54 + 36 + 53 tonnes.
Length over couplings: 26.15 + 26.40 + 26.15 m.
Max. Speed: 125 km/h.
Max. Speed: 125 km/h.

Disc brakes. Automatic doors. PA. 2881–2880 were converted from plan Y1 to plan Y2 by the addition of the trailer car and the declassification of the AB (The "first class" seats were similar to the second class ones in any case). 2881–2895 are plan Y3.

2836 (2036)	2846 (2046)	2856 (2056)	2866 (2066)	2876 (2076)	2886 *
2837 (2037)	2847 (2047)	2857 (2057)	2867 (2067)	2877 (2077)	2887 *
2838 (2038)	2848 (2048)	2858 (2058)	2868 (2068)	2878 (2078)	2888 *
2839 (2039)	2849 (2049)	2859 (2059)	2869 (2069)	2879 (2079)	2889 *
2840 (2040)	2850 (2050)	2860 (2060)	2870 (2070)	2880 (2080)	2890 *
2841 (2041)	2851 (2051)	2861 (2061)	2871 (2071)	2881 *	2891 *
2842 (2042)	2852 (2052)	2862 (2062)	2872 (2072)	2882 *	2892 *
2843 (2043)	2853 (2053)	2863 (2063)	2873 (2073)	2883 *	2893 *
2844 (2044)	2854 (2054)	2864 (2064)	2874 (2074)	2884 *	2894 *
2845 (2045)	2855 (2055)	2865 (2065)	2875 (2075)	2885 *	2895 *

PLAN mP SINGLE UNIT POSTAL VANS

The crimson and yellow postal units, formerly painted dark red are unique amongst NS units in having conventional drawgear and buffers to facilitate the taking of a trailing load of up to five special four-wheeled postal vans. They can be seen stabled at most of the principal centres during daylight hours as most of their workings occur at night time. Owned by the PTT (Post Office). At weekends they are stabled as follows:

Amsterdam (1), Arnhem (4), Den Bosch (2), Den Haag HS (2), Groningen (2), Haarlem (2), Leeuwarden (2), Roosendaal (2), Rotterdam (3), Sittard (2), Utrecht (4), Zwolle (3).

▲Plan Y2 'Sprinter' EMU No. 2875 runs into Hoofddorp with the 15.41 Amsterdam RAI–Den Haag CS stopping service on 01/08/88. *David Brown*

▼Postal unit No. 3032 at Den Haag HS on 01/08/88. *Alex Dasi-Sutton*

Built: 1965–66.
Builder-Mech. Parts: Werkspoor.
Builder-Elec. Parts: Smit.
Wheel Arrangement: Bo–Bo.
Traction Motors: 4 x Heemaf 145 kW.
Weight: 54 tonnes.
Length over couplings: 26.40 m.
Max. Speed: 140 km/h.

3001	3006	3011	3015	3019	3024	3028	3032
3002	3007	3012	3016	3020	3025	3029	3033
3003	3008	3013	3017	3022	3026	3030	3034
3004	3009	3014	3018	3023	3027	3031	3035
3005							

PLAN Z0/Z1 (ICM1/2) 3-CAR UNITS

These striking modern Intercity units, also known as IC3 units, have raised cabs and through gangways. The coaches in these sets formed the basis for the new loco-hauled ICR stock now familiar over many NS routes. The prototype units (4001–7) have been modified to standard as far as the basic layout is concerned, but they still retain different interior decor. Later units have less seats in the driving trailer, to allow for bicycle storage, and earlier units will be modified retrospectively. The units can be found on the Amsterdam CS/Den Haag/ Rotterdam CS–Groningen/Leeuwarden and the Amsterdam/Den Haag–Enschede services, but because of traffic growth certain trains are still formed of 1954 stock.

BD+AB+B (DMSO (pantry)–TCsoL–DTSOL).
Built: 1977*/83–89.
Builder-Mech. Parts: Talbot.
Builder-Elec. Parts: Oerlikon.
Wheel Arrangement: Bo–Bo+2–2+2–2.
Traction Motors: 4 x Oerlikon 312 kW.
Accommodation: 54S + 35F 31S 1L + 63S 1L.
Weight: 59 + 42 + 42 tonnes.
Length over couplings: 27.05 + 26.50 + 27.05 m.

Disc brakes. Automatic doors. PA.
Special Livery: **S** Aegon advertising livery (white/grey/blue).

4001 **I** *	4016 **I**	4028 **I**	4040 **I**	4052 **I**	4064 **I**	4076 **I**	4087 **I**
4002 **I** *	4017 **I**	4029 **I**	4041 **I**	4053 **I**	4065 **I**	4077 **I**	4088 **I**
4003 **I** *	4018 **I**	4030 **I**	4042 **I**	4054 **I**	4066 **I**	4078 **I**	4089 **I**
4004 **I** *	4019 **I**	4031 **I**	4043 **I**	4055 **I**	4067 **I**	4079 **I**	4090 **I**
4005 **I** *	4020 **I**	4032 **I**	4044 **I**	4056 **I**	4068 **I**	4080 **I**	4091 **I**
4006 **I** *	4021 **I**	4033 **I**	4045 **I**	4057 **I**	4069 **I**	4081 **I**	4092 **I**
4007 **I** *	4022 **I**	4034 **I**	4046 **I**	4058 **I**	4070 **I**	4082 **I**	4093 **I**
4011 **S**	4023 **I**	4035 **I**	4047 **I**	4059 **I**	4071 **I**	4083 **I**	4094 **I**
4012 **S**	4024 **S**	4036 **I**	4048 **I**	4060 **I**	4072 **I**	4084 **I**	4095 **I**
4013 **I**	4025 **I**	4037 **I**	4049 **I**	4061 **I**	4073 **I**	4085 **I**	4096 **I**
4014 **I**	4026 **I**	4038 **I**	4050 **I**	4062 **I**	4074 **I**	4086 **I**	4097 **I**
4015 **I**	4027 **I**	4039 **I**	4051 **I**	4063 **I**	4075 **I**		

PLAN Z2 (ICM3) 4-CAR UNITS

These units are on order, and are a four-car version of the IC3s. They are for delivery commencing 1990.

BD+B+A+B (DMSO (pantry)–TSOL–TFsoL–DTSOL).
Built: 1990 onwards.
Builder-Mech. Parts: Talbot.
Builder-Elec. Parts: Oerlikon.
Wheel Arrangement: Bo–Bo+Bo–2+2–2+2–2.
Traction Motors: 6 x Oerlikon 312 kW.
Accommodation: 54S + 88S 1L + 59F 1L + 72S.
Weight: 59 + 50 + 42 + 42 tonnes.
Length over couplings: 27.05 + 26.50 + 26.50 + 27.05 m.

Max. Speed: 160 km/h.

Disc brakes. Automatic doors. PA.

4201	4205	4209	4213	4217	4221	4225	4228
4202	4206	4210	4214	4218	4222	4226	4229
4203	4207	4211	4215	4219	4223	4227	4230
4204	4208	4212	4216	4220	4224		

TRAMS

The **Sneltram Utrecht–Nieuwegein/Ijsselstein** is operated by the West Nederland Bus Company but the vehicles are owned by the NS.

SIX-AXLE ARTICULATED TRAMS

Built: 1981–83.
Builder-Mech. Parts: SIG, Neuhausen, Switzerland.
Builder-Elec. Parts: BBC.
Wheel Arrangement: B–2–B.
Traction Motors: 2 x BBC 228 kW.
Accommodation: 98.
Weight: tonnes.
Length over couplings: 29.80 m.
Max. Speed: 80 km/h.

5001	5005	5009	5013	5016	5019	5022	5025
5002	5006	5010	5014	5017	5020	5023	5026
5003	5007	5011	5015	5018	5021	5024	5027
5004	5008	5012					

Sneltram No. 5018 at Achterveld (Ijsselstein) bound for Utrecht on 31/07/88.
Alex Dasi-Sutton

Plan W2-B11 open second No. 50 84 21-37 523-1 at the rear of the 11.31 Maastricht–Zandvoort-an-See service at Roermond on 27/07/88.
David Brown

HAULED COACHING STOCK

GENERAL

Locomotive hauled coaching stock on the NS is generally referred to by its "plan", which combines the information which in Britain would be provided by the mark and the lot number. The NS type code consists of the "Plan" plus the type code as described in the introductory section of this book.

After A or B comes the number of compartments (or windows in open stock) for each class type, e.g. A4B6–four first class and six second class compartments.

Four categories of stock are in service:

(a) **Internal stock.** As its title suggests, this stock may only work on NS lines. Train heating is 1000 V 50 Hz and 1500 V dc only.
(b) **Buurland stock.** This stock may work into adjacent countries, i.e. West Germany and Belgium. Train heating–all voltages.
(c) **Benelux Stock.** This is similar to Buurland stock, but is fitted with jumper cables for push-pull working.
(d) **International stock.** May work anvwhere. Train heating–all voltages.

NUMBERING SYSTEM

Loco-hauled coaches are not listed in strict numerical order. It will be seen that the serial numbers of ICR and plan W stock have been chosen so that they do not duplicate one another, although they do duplicate the new double-deckers. Coaches are therefore listed in three series, Plan ICR and W, Double-deckers and Miscellaneous.

ICR & PLAN W STOCK

ICR is the name given to the new NS Inter-City stock. It comes in three varieties: Internal, Buurland and Benelux. The internal stock has fixed steps, wheras the Buurland and Benelux stock have folding steps. The Benelux stock has jumper cables for push-pull working. All ICR stock has power-operated doors, pressure ventilation, disc brakes and PA. Plan W coaches have been refurbished for use with ICR stock. Plan W1 were used in the original Benelux push-pull sets, whereas plan W2 were for internal use only. All plan W vehicles are now classified as internal .

ICR and W coaches are listed in order of serial number as these do not duplicate.

PLAN ICR3-A4B6 — OPEN COMPOSITE

Built: 1986. Benelux push-pull stock.
Builder: Talbot.
Length over Buffers: 26.4 m.
UIC Prefix: 50 84 30-70
Seats: 23F 48S 2L.
Weight: 40 tonnes.

061-9 **B**
062-7 **B**
063-5 **B**
064-3 **B**
065-0 **B**
066-8 **B**
067-6 **B**
068-4 **B**
071-8 **B**
072-6 **B**

PLAN ICR3-Bvk — DRIVING OPEN SECOND

Built: 1986. Benelux push-pull stock.
Builder: Talbot.
Length over Buffers: 26.4 m.
UIC Prefix: 50 84 28-70
Seats: 64S 1L.
Weight: 41 tonnes.

101-7 **B**
102-5 **B**
103-3 **B**
104-1 **B**
105-8 **B**
106-6 **B**
107-4 **B**
108-2 **B**
111-6 **B**
112-4 **B**
113-2 **B**

PLAN ICR3-B10 — OPEN SECOND

Built: 1986–87. Benelux push-pull stock.
Builder: Talbot.
Length over Buffers: 26.4 m.
UIC Prefix: 50 84 20-70
Seats: 80S 1L
Weight: 40 tonnes.

401-9 **B**	404-3 **B**	407-6 **B**	411-8 **B**	413-4 **B**	415-9 **B**	417-5 **B**	421-7 **B**
402-7 **B**	405-0 **B**	408-4 **B**	412-6 **B**	414-2 **B**	416-7 **B**	418-3 **B**	422-5 **B**
403-5 **B**	406-8 **B**						

PLAN ICR4-B10 OPEN SECOND

Built: 1988. Benelux push-pull stock. To be renumbered from 581–5.
Builder: Talbot. **Seats:** 80S 1L
Length over Buffers: 26.4 m. **Weight:** 40 tonnes.
UIC Prefix: 50 84 20-70

423-3 | 424-1 | 425-8 | 426-6 | 427-4

PLAN ICR4-B10 OPEN SECOND

Built: 1988. Benelux push-pull stock. To be converted from open firsts in the series 561–4.
Builder: Talbot. **Seats:** 80S 1L
Length over Buffers: 26.4 m. **Weight:** 40 tonnes.
UIC Prefix: 50 84 20-70

431-6 | 432-4 | 433-2

PLAN W1-B11 OPEN SECOND

Original Benelux push & pull stock. Swivel doors. PA. Renumbered from 50 84 21-30 series.

Built: 1966–67.
Builder: Werkspoor. **Seats:** 88S 1L.
Length over Buffers: 26.4 m. **Weight:** 35 tonnes.
UIC Prefix: 50 84 21-37

451-2 **I**	454-6 **I**	457-9 **I**	462-9 **I**	465-2 **I**	468-6 **I**	473-6 **I**	476-9 **I**
452-0 **I**	455-3 **I**	458-7 **I**	463-7 **I**	466-0 **I**	471-0 **I**	474-4 **I**	477-7 **I**
453-8 **I**	456-1 **I**	461-1 **I**	464-5 **I**	467-8 **I**	472-8 **I**	475-1 **I**	

PLAN ICR3-A SEMI-OPEN FIRST

Built: 1986. Benelux push-pull stock.
Builder: Talbot. **Seats:** 59F 1L. (24F in compts, 35F open).
Length over Buffers: 26.4 m. **Weight:** 41 tonnes.
UIC Prefix: 50 84 10-70

481-3 **B**	483-9 **B**	485-4 **B**	486-2 **B**	487-0 **B**	488-8 **B**	491-2 **B**	492-0 **B**
482-1 **B**	484-7 **B**						

PLAN W2-B11 OPEN SECOND

Internal stock. Swivel doors. PA.

Built: 1968.
Builder: Werkspoor. **Seats:** 88S 1L.
Length over Buffers: 26.4 m. **Weight:** 35 tonnes.
UIC Prefix: 50 84 21-37

501-7 **I**	505-8 **I**	511-6 **I**	514-0 **I**	517-3 **I**	522-3 **I**	525-6 **I**	528-0 **I**
502-5 **I**	506-6 **I**	512-4 **I**	515-7 **I**	518-1 **I**	523-1 **I**	526-4 **I**	531-4 **I**
503-3 **I**	507-4 **I**	513-2 **I**	516-5 **I**	521-5 **I**	524-9 **I**	527-2 **I**	532-2 **I**
504-1 **I**	508-2 **I**						

PLAN ICR4-A10 SEMI-OPEN FIRST

Built: 1988. Buurland Stock. Through-wired for Benelux service if required.
Builder: Talbot. **Seats:** 59F 1L. (24F in compts, 35F open).
Length over Buffers: 26.4 m. **Weight:** 41 tonnes.
UIC Prefix: 50 84 10-70

561-2 **I** | 562-0 **I** | 563-8 **I** | 564-6 **I**

PLAN ICR4-B10 OPEN SECOND

Built: 1988. Buurland Stock. Through-wired for Benelux service if required.
Builder: Talbot. **Seats**: 80S 1L
Length over Buffers: 26.4 m. **Weight**: 40 tonnes.
UIC Prefix: 50 84 20-70

581-8 I	582-6 I	583-4 I	584-2 I	585-9 I

PLAN ICR4-BKD KITCHEN BRAKE OPEN 2nd

These vehicles consist of a brake open second plus a kitchenette from which a trolley refreshment service operates.

Built: 1988. Buurland Stock. Through-wired for Benelux service if required.
Builder: Talbot. **Seats**: 49S 1L.
Length over Buffers: 26.4 m. **Weight**: 41 tonnes.
UIC Prefix: 50 84 82-70

591-2 I	592-0 I	593-8 I

PLAN ICR2-A10 SEMI-OPEN FIRST

Built: 1982–84. Internal Stock.
Builder: Talbot. **Seats**: 59F 1L. (24F in compts, 35F open).
Length over Buffers: 26.4 m. **Weight**: 41 tonnes.
UIC Prefix: 50 84 10-77

601-9 I	605-0 I	611-8 I	615-9 I	621-7 I	625-8 I	631-6 I	635-7 I
602-7 I	606-8 I	612-6 I	616-7 I	622-5 I	626-6 I	632-4 I	636-5 I
603-5 I	607-6 I	613-4 I	617-5 I	623-3 I	627-4 I	633-2 I	637-3 I
604-3 I	608-4 I	614-2 I	618-3 I	624-1 I	628-2 I	634-0 I	638-1 I

ICR-A10 SEMI-OPEN FIRST

Built: 1981–88. Buurland Stock.
Builder: Talbot. **Seats**: 59F 1L. (24F in compts, 35F open).
Length over Buffers: 26.4 m. **Weight**: 41 tonnes.
UIC Prefix: 50 84 10-70

641–648. Plan ICR4. 1988.
651–687. Plan ICR1. 1981–83.
688–691. Plan ICR4. 1988.

641-2 I	647-9 I	654-5 I	661-0 I	666-9 I	673-5 I	678-4 I	685-9 I
642-0 I	648-7 I	655-2 I	662-8 I	667-7 I	674-3 I	681-8 I	686-7 I
643-8 I	651-1 I	656-0 I	663-6 I	668-5 I	675-0 I	682-6 I	687-5 I
644-6 I	652-9 I	657-8 I	664-4 I	671-9 I	676-8 I	683-4 I	688-3 I
645-3 I	653-7 I	658-6 I	665-1 I	672-7 I	677-6 I	684-2 I	691-7 I
646-1 I							

PLAN ICR2-B10 OPEN SECOND

Built: 1982–84. Internal Stock.
Builder: Talbot. **Seats**: 80S 1L
Length over Buffers: 26.4 m. **Weight**: 40 tonnes.
UIC Prefix: 50 84 20-77

701-5 I	711-4 I	721-3 I	731-2 I	741-1 I	748-6 I	757-7 I	766-8 I
702-3 I	712-2 I	722-1 I	732-0 I	742-9 I	751-0 I	758-5 I	767-6 I
703-1 I	713-0 I	723-9 I	733-8 I	743-7 I	752-8 I	761-9 I	768-4 I
704-9 I	714-8 I	724-7 I	734-6 I	744-5 I	753-6 I	762-7 I	771-8 I
705-6 I	715-5 I	725-4 I	735-3 I	745-2 I	754-4 I	763-5 I	772-6 I
706-4 I	716-3 I	726-2 I	736-1 I	746-0 I	755-1 I	764-3 I	773-4 I
707-2 I	717-1 I	727-0 I	737-9 I	747-8 I	756-9 I	765-0 I	774-2 I
708-0 I	718-9 I	728-8 I	738-7 I				

ICR-B10 OPEN SECOND

Built: 1981–88. Buurland Stock.
Builder: Talbot.
Length over Buffers: 26.4 m.
UIC Prefix: 50 84 20-70

Seats: 80S 1L
Weight: 40 tonnes.

801–874. Plan ICR1. 1981–82.
875–881. Plan ICR4. 1988.

801-0 **I**	812-7 **I**	822-6 **I**	832-5 **I**	842-4 **I**	852-3 **I**	862-2 **I**	872-1 **I**
802-8 **I**	813-5 **I**	823-4 **I**	833-3 **I**	843-2 **I**	853-1 **I**	863-0 **I**	873-9 **I**
803-6 **I**	814-3 **I**	824-2 **I**	834-1 **I**	844-0 **I**	854-9 **I**	864-8 **I**	874-7 **I**
804-4 **I**	815-0 **I**	825-9 **I**	835-8 **I**	845-7 **I**	855-6 **I**	865-5 **I**	875-4 **I**
805-1 **I**	816-8 **I**	826-7 **I**	836-6 **I**	846-5 **I**	856-4 **I**	866-3 **I**	876-2 **I**
806-9 **I**	817-6 **I**	827-5 **I**	837-4 **I**	847-3 **I**	857-2 **I**	867-1 **I**	877-0 **I**
807-7 **I**	818-4 **I**	828-3 **I**	838-2 **I**	848-1 **I**	858-0 **I**	868-9 **I**	878-8 **I**
808-5 **I**	821-8 **I**	831-7 **I**	841-6 **I**	851-5 **I**	861-4 **I**	871-3 **I**	881-2 **I**
811-9 **I**							

ICR-BKD KITCHEN BRAKE OPEN SECOND

Buurland stock. These vehicles consist of a brake open second plus a kitchenette from which a trolley refreshment service operates.

Built: 1981–88.
Builder: Talbot.
Length over Buffers: 26.4 m.
UIC Prefix: 50 84 82-70

Seats: 53S 1L. (49S 1L ICR4)
Weight: 41 tonnes.

901–953. Plan ICR1. 1981–82.
954–958. Plan ICR4. 1988.

901-3 **I**	907-0 **I**	915-3 **I**	923-7 **I**	931-0 **I**	937-7 **I**	945-0 **I**	953-4 **I**
902-1 **I**	908-8 **I**	916-1 **I**	924-5 **I**	932-8 **I**	938-5 **I**	946-8 **I**	954-2 **I**
903-9 **I**	911-2 **I**	917-9 **I**	925-2 **I**	933-6 **I**	941-9 **I**	947-6 **I**	955-9 **I**
904-7 **I**	912-0 **I**	918-7 **I**	926-0 **I**	934-4 **I**	942-7 **I**	948-4 **I**	956-7 **I**
905-4 **I**	913-8 **I**	921-1 **I**	927-8 **I**	935-1 **I**	943-5 **I**	951-8 **I**	957-5 **I**
906-2 **I**	914-6 **I**	922-9 **I**	928-6 **I**	936-9 **I**	944-3 **I**	952-6 **I**	958-3 **I**

PLAN ICR3-BKD KITCHEN BRAKE OPEN 2nd

These vehicles consist of a brake open second plus a kitchenette from which a trolley refreshment service operates.

Built: 1986. Benelux push-pull stock.
Builder: Talbot.
Length over Buffers: 26.4 m.
UIC Prefix: 50 84 82-70

Seats: 45S 1L.
Weight: 41 tonnes.

971-6 **B**	973-2 **B**	975-7 **B**	976-5 **B**	977-3 **B**	978-1 **B**	981-5 **B**	982-3 **B**
972-4 **B**	974-0 **B**						

DOUBLE-DECKER STOCK

The new double-decker stock operates on outer suburban trains on the Amsterdam CS–Den Helder, Amsterdam–Lelystad, Den Haag–Roosendaal–Tilburg and Rotterdam–Amersfoort routes. The Driving coaches are named after endangered species.

DOUBLE DECKER DRIVING OPEN SECOND (Bvs)

Built: 1985–86.
Builder: Talbot.
Length over Buffers: 26.89 m.
Seats: 108S 1L. (64 upper deck, 44 lower deck).
Weight: tonnes.
UIC Prefix: 50 84 26-37

101-1 Kondor	106-0 Arend	112-8 Cheeta
102-9 Ooieveer	107-8 Zeehond	113-6 Dolfijn
103-7 Bison	108-6 Olifant	114-4 Otter
104-5 Walvis	111-0 Tijger	115-1 Panda
105-2 Neushoorn		

DOUBLE DECKER SECOND (Bv)

Built: 1985–86.
Builder: Talbot.
Length over Buffers: 26.4 m.
Accommodation: 140S 1L. (64 upper deck, 76 lower deck).
Weight: tonnes.
UIC Prefix: 50 84 26-37

401-5	406-4	412-2	416-3	422-1	426-2	432-0	436-1
402-3	407-2	413-0	417-1	423-9	427-0	433-8	437-9
403-1	408-0	414-8	418-9	424-7	428-8	434-6	438-7
404-9	411-4	415-5	421-3	425-4	431-2	435-3	441-1
405-6							

DOUBLE DECKER COMPOSITE (ABv)

Built: 1985–86.
Builder: Talbot.
Length over Buffers: 26.4 m.
Accommodation: 64F 60S 1L. (32F 24S upper deck, 32F 36S lower deck).
Weight: tonnes.

UIC Prefix: 50 84 26-37

601-0	605-1	611-9	615-0	621-8	625-9	628-3	633-3
602-8	606-9	612-7	616-8	622-6	626-7	631-7	634-1
603-6	607-7	613-5	617-6	623-4	627-5	632-5	635-8
604-4	608-5	614-3	618-4	624-2			

MISCELLANEOUS STOCK

PLAN L-DF BICYCLE BRAKE

Built: 1958 as post office vehicles. Converted 1983 to carry 30 cycles.
Builder: Werkspoor.
Length over Buffers: 23.05 m.
Accommodation: None.
Weight: 45 tonnes.

50 84 92-37 001-2	50 84 92-37 002-0	50 84 92-37 003-8	50 84 92-37 004-6

TYPE SR BUFFET & DANCING CAR

Built: 1953–54. International stock. Used on holiday trains and Motorails.
Builder: Length over Buffers: 26.4 m.
Accommodation:
Weight: 42 tonnes.

51 84 88-80 001-3 I	51 84 88-40 003-8 I	51 84 88-80 005-4 I	51 84 88-40 007-9 I
51 84 88-80 002-1 I	51 84 88-80 004-7 I	51 84 88-40 006-1 I	

TYPE WR RESTAURANT CAR

International stock. Built originally for the DB and sold to "Apfelpfeil" for holiday traffic. Bought by the NS when Apfelpfeil went bankrupt in 1980.

Built: 1954–55.
Builder:
Length over Buffers: 26.4 m.
Accommodation: 30U.
Weight: 42 tonnes.

61 84 88-80 011-0 I	61 84 88-40 012-7 I	61 84 88-80 013-6 I	61 84 88-80 014-4 I

ROYAL SALOONS

Originally composites. Rebuilt as Royal saloons 1955/53. ...001 was built for Queen Juliana and consists of a toilet, bathroom, bedroom, full width lounge, coupé, kitchen and attendant's compartment with boiler. ...002 was built for Prince Bernhard and Princess Beatrix and consists of two bedrooms, two bathrooms, a combined sitting room and bedroom, an attendant's compartment with boiler and a governess' compartment.

Built: 1932.
Builder: Werkspoor.
Length over Buffers: 21.80 m.
Weight: 56 (53*) tonnes.

61 84 89-30 001-1 (9)
61 84 89-30 002-9 (8)*

ROYAL STAFF COACH

Originally a plan K composite.

50 84 38-37 082-9

SLEEPING CAR TYPE Mü

Built: 1973.
Builder: Fiat.
Length over Buffers: 28.40 m.
Berths: 36. Convertible first/second.
Weight: 55 tonnes.

71 84 72-80 622-0

SLEEPING CAR TYPE T2S

Built: 1975.
Builder:
Length over Buffers: 26.4 m.
Berths: 34. (17 twin berth compartments).
Weight: 55 tonnes.

71 84 75-70 458-8
71 84 75-70 459-6

PLAN E STOCK

This was internal stock of low profile. Hand-operated sliding doors. Dark blue livery. The following vehicles are in museum stock and have been reinstated for use with steam and other special trains for the NS 150 celebrations.

PLAN E-B9 OPEN THIRD

Built: 1968.
Builder: Werkspoor.
Length over Buffers: 23.05 m.
Accommodation: 72T 2L.
Weight: 45 tonnes.

6669 (50 84 29-37 373-3)
6625 (50 84 29-37 374-1)
6703 (50 84 29-37 352-7)
6712 (50 84 29-37 368-3)
6714 (50 84 29-37 371-7)

PLAN E-RD BRAKE RESTAURANT CAR

Built: 1955–56.
Builder: Beijnes.
Length over Buffers: 23.05 m.
Accommodation: 22 buffet.
Weight: 46 tonnes.

6925 (50 84 87-37 231-2)

DEPARTMENTAL VEHICLES

80 84 978-1 002-4 Ultrasonic test unit (diesel hydraulic)
80 84 978-1 505-6 three-phase test car
80 84 978-1 601-3 mDw ATB test car ('Jules')

NS PRESERVED LOCOMOTIVES AND MULTIPLE UNITS

STEAM LOCOMOTIVES:

Number	Wheels	Built	Status	Location
13	2–4–0	1865	M	Utrecht
89	2–4–0	1880	M	Utrecht
107	4–4–0	1889	M	Utrecht
326	2–4–0	1881	M	Utrecht
2104	4–4–0	1914	M	Utrecht
3737	4–6–0	1911	MA	Utrecht
5085	2–10–0	1945	M	Utrecht
6317	4–8–4T	1931	M	Utrecht
7742	0–6–0T	1914	MA	SHM
8811	0–6–0ST	1943	M	SSN
8815	0–6–0ST	1943	MR	UK
8817	0–6–0ST	1943	P	Autotron Rosmalen
8826	0–6–0ST	1943	P	Tilburg
657	0–4–0T	1901	MR	MBS

DIESEL & ELECTRIC LOCOMOTIVES:

Number	Wheels	Built	Status	Location
106	B pm	1931	MR	VSM Beekbergen
137	B pm	1929	M	Utrecht
451	C de	1956	MA	MBS
311	B de	1940	M	Utrecht
508	C de	1944	M	Utrecht
532	C de	1954	MA	VSM
1010	1A–Bo–A1 e	1949	M	STIBANS. Roosendaal
1125	Bo–Bo e	1951	M	Utrecht. Restored as 1122
1501	Co–Co e	1954	MA	Feijenoord
1502	Co–Co e	1954	MA	Utrecht. Now BR 27000
1505	Co–Co e	1954	M	Greater Manchester, UK
2498	Bo–Bo de	1955	M	Utrecht
197	B pm	1932	M	MBS
1xx	B pm	193	M	MBS

DIESEL & ELECTRIC RAILCARS:

Number	Wheels	Built	Status	Location
27	3-car DEMU	1934	M	Utrecht
41	Bo–Bo DMU	1935	M	Roosendaal
252	2-car EMU	1950	MS	STIBANS. Roosendaal
273	4-car EMU	1952	MS	Roosendaal
9107	Bo–Bo EMU	1924	MA	Utrecht. ('De Blokkendoos' [box of bricks]).
8104	EMU dt	1924	MA	Utrecht. (Driving Trailer for above).

Ex DB/DR LOCOMOTIVES PRESERVED IN HOLLAND:

Number	Wheels	Built	Status	Location
01.107	4–6–2	1937	MR	SSN
23.023	2–6–2	1952	MA	SSN
23.071	2–6–2	1956	MA	VSM
23.076	2–6–2	1956	MA	VSM
41.105	2–8–2	1939	MA	SSN
50.125	2–10–0	1941	MR	SSN
52.3879	2–10–0	1944	MA	SSN
52.8139	2–10–0	1944	M	VSM
65.018	2–8–4T	1956	MA	SSN
94.640	0–10–0T	1923	P	Gennep

3.LUXEMBOURG RAILWAYS (CFL)

The Luxembourg Railways are known by the abbreviation CFL (Societé Nationale des Chemins de Fer Luxembourgeois). The total length of the system is only 270 km, but this does not mean that the network is uninteresting. The CFL operates locomotives and multiple units of types that are also found in neighbouring countries, and the city of Luxembourg also sees through workings of locomotives from the DB, SNCB and SNCF.

NUMBERING SYSTEM

The CFL loco numbering system is based on the horse power of the locomotives. 1801 is an 1800 hp locomotive, whilst 3620 is one of 3600 hp. However, departmental vehicles are numbered 1XXX whilst the new Z2 EMUs are numbered 2XXX.

DEPOTS AND WORKSHOPS

With such a compact system there is no need for many depots. All locomotives are allocated to Luxembourg shed. There is a sub-shed at Wasserbillig and there are stabling points at Bettembourg Yard, Pétange and Esch sur Alzette. Locomotives also stable overnight at the various branch termini as required. Because of the small fleet only one workshop is required and this is located opposite the station at Luxembourg.

DEVELOPMENTS

The next few years will see some changes to the CFL fleet. The line from Luxembourg to Gouvy is to be electrified. There will thus be a requirement for more electric locomotives and less diesels. The CFL diesel and electric fleet is ageing and in need of renewal. Time is definitely running out for the existing fleet of locos and DMUs so do not leave it too late to visit this interesting system.

DIESEL MULTIPLE UNITS

200 CLASS 2-CAR UNITS

These vehicles were originally numbered 201A, 201B etc. It is pleasing to note that the CFL has something numbered 208! Note that on DMUs the numbers refer to the car, whereas on EMUs the numbers refer to the complete unit.

B+B (2DMSOL).

Built: 1956.
Builder: Westwaggon.
Wheel Arrangement: 1A–2 + 2–A1.
Engine: Deutz A121614 of 149 kW (one per car).
Transmission: Hydraulic. Voith.
Accommodation: 94S 1L + 94S 1L.
Total Weight: 31.5 + 31.5 tonnes.
Length over Couplings: 23.72 + 23.72 m.
Max. Speed: 105 km/h.

201	211	203	213	205	215	207	217
202	212	204	214	206	216	208	218

ELECTRIC MULTIPLE UNITS

250 CLASS 2-CAR UNITS

25 kV ac. These are based on similar units supplied to the SNCF and feature monomotor bogies and thyristor control. Up to three units may work in multiple.

AB+BD (DMCOL–DMBSO).

Built: 1975.
Builder-Mech. Parts: CF.
Builder-Elec. Parts: MTE, CEM.
Wheel Arrangement: 2–B+2–2.
Traction Motors: 2 x 308 kW.
Accommodation: 14F 80S 1L + 74S.
Total Weight: 79.5 tonnes.
Length over couplings: 25.325 + 25.325 m.
Max. Speed: 120 km/h.

251 | 252 | 253 | 254 | 255 | 256

260 CLASS 3-CAR UNITS

25 kV ac. These units have been acquired second-hand from the SNCF. They are 3-car versions of class 250 and have the same features.

BD+B+AB (DMBSO–TSO–DTCOL).

Built: 1965.
Builder-Mech. Parts: CF.
Builder-Elec. Parts: CEM.
Wheel Arrangement: 2–B+2–2+2–2.
Traction Motors: 2 x 308 kW.
Accommodation: 79S + 107S + 36F 51S 1L.
Total Weight: 90 tonnes.
Length over couplings: 74.450 m.
Max. Speed: 120 km/h.

261 | 262

2000 CLASS 2-CAR UNITS

To be introduced 1990, these EMUs are similar to the SNCF Z2 series.

ABD+B (DMBCOL–DTSOL).

Built: 1990.
Builders: De Dietrich.
Wheel Arrangement: Bo–Bo+2–2.
Traction Motors: 4 x 305 kW.
Seats: 24F 43S 1L + 84S 1L.
Weight: + tonnes.
Length over couplings: 25.10 + 25.10 m.
Max. Speed: 160 km/h.

2001 | 2004 | 2007 | 2010 | 2013 | 2016 | 2019 | 2021
2002 | 2005 | 2008 | 2011 | 2014 | 2017 | 2020 | 2022
2003 | 2006 | 2009 | 2012 | 2015 | 2018

DIESEL LOCOMOTIVES

Note: Self-propelled departmental vehicles are also numbered in this series.

800 CLASS Bo–Bo

These locos are unique to the CFL and are typical American switches of the early 1950s. They are used for shunting and trip working.

Built: 1954.
Builder: AFB.
Engine: GM 8-567B of 600 kW at 835 rpm.
Transmission: Electric. Four GM axle-hung traction motors.
Train Heating: None
Maximum Tractive Effort: 179 kN.
Driving Wheel Dia.: 1050 mm.
Weight in Full Working Order: 74 tonnes.
Length over Buffers: 13.795 m.
Max. Speed: 80 km/h.

801 | 802 | 803 | 804 | 805 | 806

▲CFL diesels 802 and 903 stand at Luxembourg depot on 26/05/89. *Graham Scott-Lowe*

▼1002 is seen at Luxembourg depot. *Brian Garvin*

850 CLASS Bo–Bo

Mixed traffic use.

Built: 1956–57.
Builder: BL.
Engine: SACM MGO V12SH of 615 kW at ? rpm.
Transmission: Electric. Four BL axle-hung traction motors.
Train Heating: None.
Maximum Tractive Effort: 174 kN.
Driving Wheel Dia.: 1100 mm.
Weight in Full Working Order: 72 tonnes.
Length over Buffers: 14.75 m.
Max. Speed: 105 km/h.

851 | 852 | 853 | 854 | 855 | 856 | 857 | 858

900 CLASS Bo–Bo

Mixed traffic use.

Built: 1958.
Builder: BL.
Engine: SACM MGO V12SHR of 690 kW at ? rpm.
Transmission: Electric. Four BL axle-hung traction motors.
Train Heating: None.
Maximum Tractive Effort: 174 kN.
Driving Wheel Dia.: 1100 mm.
Weight in Full Working Order: 72 tonnes.
Length over Buffers: 14.75 m.
Max. Speed: 105 km/h.

902 is named STEINFURT.

901 | 903 | 905 | 907 | 909 | 911 | 912 | 913
902 | 904 | 906 | 908 | 910

1000 CLASS B

Used for light shunting duties and usually found at Bettembourg p.w. Depot, Pétange Wagon Works, Luxembourg and Ettelbruck.

Built: 1972.
Builder: Jung.
Engine: Deutz F12 L413 of 186 kW.
Transmission: Hydraulic. Voith.
Train Heating: None
Maximum Tractive Effort: 94 kN.
Driving Wheel Dia.: 950 mm.
Weight in Full Working Order: 32 tonnes.
Length over Buffers: 7.20 m.
Max. Speed: 60 km/h.

1001 | 1002 | 1003 | 1004

1010 CLASS B

This shunter is usually found at Pétange Works.

Built: 1964.
Builder: Henschel.
Engine: Henschel 6R 1215A of 160 kW.
Transmission: Hydraulic. Voith.
Train Heating: None
Maximum Tractive Effort: 64 kN.
Driving Wheel Dia.: 850 mm.
Weight in Full Working Order: 22 tonnes.
Length over Buffers: 7.10 m.
Max. Speed: 24 km/h.

1011 (2011)

1020 CLASS B

Used for light shunting duties, usually 1 at Luxembourg shed, 1 at Luxembourg Works and 2 spare. 1023/4 are fitted with auto-couplers.

Built: 1952–57.
Builder: Deutz.
Engine: Deutz A8 L614 of 100 kW.
Transmission: Hydraulic. Voith.
Train Heating: None
Weight in Full Working Order: 22 tonnes.

▲**Overhead line unit No. 1061 is seen on wiring duties at Luxembourg station.** *Brian Garvin*

▼**3614 waits at Luxembourg with the 17.30 to Wasserbillig on 26/05/89. The first coach is one which has been upgraded to 140 km/h and painted in the new cream & green livery.** *Graham Scott-Lowe*

Maximum Tractive Effort: 64 kN.
Driving Wheel Dia.: 850 mm.
Length over Buffers: 7.57 m.
Max. Speed: 53 km/h.

1021 | 1022 | 1023 | 1024

1030 CLASS B

New shunters, fitted with remote control.

Built: 1988.
Builder: Jenbacher Werke, Austria.
Engine: MTU 8V 183 TA12.
Transmission: Hydraulic. Voith L2r4SV2.
Train Heating: None
Maximum Tractive Effort: 117 kN.
Driving Wheel Dia.: 950 mm.
Weight in Full Working Order: 36 tonnes.
Length over Buffers: 8.55 m.
Max. Speed: 60 km/h.

1031 | 1032 | 1033

1050 CLASS P.W. TROLLEYS

These permanent way trolleys have a large cabin at one end for the driver and staff and a hydraulic arm at the other. 1051 was formerly numbered departmental 10.

Built: 1980–82.
Builder: Donelli/Geismar.
Engine: Deutz F8 L413F of 173.5 kW. at 2500 rpm.
Length over Buffers: 9.90 m.
Driving Wheel Dia.: 850 mm.
Weight in Full Working Order: 36 tonnes.
Max. Speed: 80 km/h.

1051 | 1052 | 1053 | 1054

1060 CLASS O.H.L. TROLLEYS

This is another departmental vehicle numbered in the capital stock series. It features a dummy pantograph and a working area above the crew accommodation. It is used for overhead line maintenance.

Built: 1985.
Builder: Donelli/Geismar.
Engine: Deutz F8 L413F of 173.5 kW at 2500 rpm.
Length over Buffers: 12.14 m.
Driving Wheel Dia.: 850 mm.
Weight in Full Working Order: 24 tonnes.
Max. Speed: 80 km/h.

1061

1070 CLASS O.H.L. TROLLEYS

Another departmental vehicle. No details available.

1071

1600 CLASS Co–Co

These are the same class as the SNCB class 52. 1601 has been cut up after being used as a spare parts bank for many years. They are usually used on the Luxembourg–Troisvierges local service.

Built: 1955.
Builder-Mech. Parts: AFB.
Builder-Elec. Parts: GM.
Engine: GM 16-567C of 1265 kW at 835 rpm.
Transmission: Electric. 6 axle-hung traction motors.
Train Heating: Steam. Vapor OK4616.
Maximum Tractive Effort: 245 kN.
Driving Wheel Dia.: 1010 mm.
Weight in Full Working Order: 108 tonnes.
Length over Buffers: 19.007 m.
Max. Speed: 120 km/h.

Rheostatic braking.

1602
1603
1604 FOND-DE-GRAS

1800 CLASS — Co–Co

These are the same class as the SNCB class 55.

Built: 1963.
Builder-Mech. Parts: BN.
Builder-Elec. Parts: ACEC/SEM.
Engine: GM 16-567C of 1435 kW at 835 rpm.
Transmission: Electric. Six axle-hung traction motors.
Train Heating: Steam. Vapor 4616.
Maximum Tractive Effort: 272 kN.
Driving Wheel Dia.: 1010 mm.
Weight in Full Working Order: 110 (114 s) tonnes.
Length over Buffers: 19.55 m.
Max. Speed: 120 km/h.

Rheostatic braking.
s–Steam heating fitted.

1801 s
1802 s Blankenberge
1803 s
1804 s
1805 MONDORF-LES-BAINS
1806 COMMUNE DE WALFERDANGE
1807 COMMUNE DE PETANGE
1808
1809
1810
1811 ETTELBRUCK
1812
1813
1814
1815 s KAUTENBACH 1881–1981
1816 s LAROCHETTE
1817 s
1818 s
1819 s PRINCE HENRI
1820 s Bettembourg

ELECTRIC LOCOMOTIVES

TYPE 3600 — Bo–Bo

These 25 kV ac electric locos are similar to the SNCF BB 12000 class.

Built: 1958–59.
Builder-Mech. Parts: MTE.
Builder-Elec. Parts: MTE.
Traction Motors: Four MTE.
One Hour Rating: 2650 kW.
Maximum Tractive Effort: kN.
Driving Wheel Dia.: 1250 mm.
Total Weight: 84 tonnes.
Length over Buffers: 15.20 m.
Max. Speed: 120 km/h.

3601
3602
3603
3604
3605
3606 DUDELANGE
3607 ESCH/ALZETTE
3608 LORENTZWEILER
3609
3610
3611
3612
3613 GUILLAUME LUXEMBOURG
3614 RUMELANGE
3616 DUDELANGE
3617
3618 WILTZ
3619 WASSERBILLIG
3620 REISERBAHN

HAULED COACHING STOCK

There are four types of coaching stock vehicles in use on the CFL, types B, AB and BD and ABD. The ABD vehicles are identical to the BD vehicles and the AB vehicles are identical to the B vehicles, except for the upholstery on the first class seats.

The livery of all coaches was dark green, but this has been changed to cream and green on vehicles upgraded to 140 km/h which are also allowed to work into Belgium, the Netherlands and Germany.

All vehicles were built by Wegmann and are 26.1 m. in length. They are listed in order of serial number.

B12 OPEN SECOND

Accommodation: 96S 2L.
Max Speed: 120 km/h.

50 82 22-10 331-1
50 82 22-10 333-7
50 82 22-10 334-5
50 82 22-10 335-2
50 82 22-10 336-0
50 82 22-10 337-8
50 82 22-10 338-6
50 82 22-10 339-4
50 82 22-10 340-2
50 82 22-10 341-0
50 82 22-10 342-8
50 82 22-10 343-6
50 82 22-10 346-9
50 82 22-10 347-7
50 82 22-10 348-5
50 82 22-10 349-3
50 82 22-10 350-1
50 82 22-10 351-9
50 82 22-10 352-7
50 82 22-10 353-5
50 82 22-10 354-3
50 82 22-10 355-0
50 82 22-10 359-2
50 82 22-10 361-8
50 82 22-10 368-3
50 82 22-10 369-1
50 82 22-10 370-9

A3B6D BRAKE OPEN COMPOSITE

Accommodation: 24F 48S 1L.
Max Speed: 120 km/h.

50 82 81-10 380-6
50 82 81-10 381-4

B9D BRAKE OPEN SECOND

Accommodation: 72S 2L.
Max Speed: 120 km/h.

50 82 82-10 390-4
50 82 82-10 391-2
50 82 82-10 392-0
50 82 82-10 393-8
50 82 82-10 394-6

A3B9 OPEN COMPOSITE

Accommodation: 24F 72S 2L.
Max Speed: 140 km/h.

50 82 32-40 414-7
50 82 32-40 415-4
50 82 32-40 417-0
50 82 32-40 418-8

B12 OPEN SECOND

Accommodation: 96S 2L.
Max Speed: 140 km/h.

50 82 22-40 432-1
50 82 22-40 444-6
50 82 22-40 445-3
50 82 22-40 456-0
50 82 22-40 457-8
50 82 22-40 458-6
50 82 22-40 460-2
50 82 22-40 462-8
50 82 22-40 463-6
50 82 22-40 464-4
50 82 22-40 465-1
50 82 22-40 466-9
50 82 22-40 467-7
50 82 22-40 471-9
50 82 22-40 472-7
50 82 22-40 473-5

A3B6D BRAKE OPEN COMPOSITE

Accommodation: 24F 48S 2L.
Max Speed: 140 km/h.

50 82 81-40 482-4
50 82 81-40 483-2
50 82 81-40 484-0
50 82 81-40 485-7
50 82 81-40 486-5
50 82 81-40 487-3

CFL PRESERVED LOCOMOTIVES AND MULTIPLE UNITS

Number	*Type*	*Built*	*Status*	*Location*
105	1A–A1 DMMU	1949	MR	Luxembourg
151	A–1 DMMU	1951	MA	Fond de Gras
455	C dh	1963	MS	Luxembourg
2001	B dh	1957	MS	Luxembourg
5519	2–10–0	1947	P	Bettembourg

For status codes see SNCB section.

MUSEUMS & MUSEUM LINES

Preservation in the three Benelux countries for many years was only represented by the official collections in Belgium and the Netherlands. Then came the preservation boom of the late 1960s and it has really taken off since then. Unfortunately the Benelux countries were some of the first in Europe to eliminate main line steam so that steam preservation has been mostly of industrial types or of locos obtained from other countries – notably West Germany. Like other countries on the European mainland, museum line operations tend to be at weekends only but there are notable exceptions with some lines having daily trains in the peak summer months. All railways/tramways are standard gauge (1435 mm) unless otherwise shown.

BELGIUM.

Brussels.

Musée Nationale des Chemins de Fer/Museum van de Spoorwegen. The National Railway Museum is located in Brussels Nord station, the entrance being off the booking hall. Only one loco is present, the others in the collection being stored at Leuven until suitable premises can be found. There are many models and photos etc. explaining the development of railways in Belgium. Open Monday–Friday and first Saturday of the month if it is not a holiday, 09.00–16.30.

Brussels.

Musée du Transport Urbain Bruxellois. MTUB. This tram and bus museum is located in the suburb of Woluwe St. Pierre at the old depot there. Access is by trams 39 or 44 and buses 36 or 42. Tourist services operate over part of route 44. Open on Saturdays, Sundays and holidays April–September, 14.00–18.00.
More than 40 trams/buses.

Buda–Vilvoorde–Verbrande Brug. 6 km.

Museum Stoomtrein der Twee Bruggen. MSTB. This line runs alongside the canal in Vilvoorde just north of Brussels. Access by train to Vilvoorde or tram 58. Operates each Sunday, April–October.
2 steam, 1 DMU.

Dendermonde–Puurs. 14 km.

Stoomspoorlijn Dendermonde–Puurs. SDP. A relatively new operation over a closed NMBS line. The depot is at Baasrode. Operates on Sundays in July and August, 14.00–18.00.
3 steam, 5 diesel, 2 DMUs.

Edegem

Antwerps Tram en Autobusmuseum. This is part of the AMUTRA organisation. Access by bus 32 from Antwerpen or Berchem stations or trams 7 and 15 from Antwerpen. The museum is located in the Fort V recreational park and is open weekends and holidays May to September, 14.00–18.00.
More than 25 trams and buses.

Erezee–Dochamps.9 km.

1000 mm. Tramway Touristique de l'Aisne. TTA. An interesting preserved tramway deep in the Ardennes and rather inaccessible. The operating base is Pont D'Erezee which is not on the railway network. Public transport details are not known. Oerates weekends Easter–September and daily in July and August.
4 steam.

Leuven.

Former NMBS depot and workshop. Most of the locomotives and rolling stock belonging to the National Collection are housed here. The premises are classed as a workshop and are not open to the public. Restoration work is carried out here. However, recognising the public interest in the collection, there are usually open days early in September.
19 steam, 2 diesels, 2 EMUs, 3 DMUs.

Liège

Musée des Transports en commun du pays de Liège. A collection of trams and buses. The museum is located at Rue Richard Heintz, 9 and is on bus routes 4, 26, 28 and 31. Open weekends and holidays April–September, 14.00–18.00.

Lobbes–Thuin. 5 km.

1000 mm. Association pour la Sauvegarde du Vicinal. ASVi. A tramway operation over a scenic cross country SNCV route. Operates Sundays and holidays June–September and additionally on Saturdays in July and August. First departure is 13.25 from Lobbes. The depot is believed to be at Anderlues.
1 steam and trams.

Maldegem

600 mm and 1435 mm. Stoom Centrum Maldegem. SCM. A museum has been established alongside the old station at Maldegem and a narrow gauge line has been constructed believed to be 5 km long. Traction engines and agricultural equipment are also present. Open Saturdays and Sundays from Easter to the end of September 09.00–17.00.
4 steam, 5 diesel (600 mm), 9 steam, 7 diesel (1435 mm).

Mariembourg–Chimay/Treignes. 30 km.

Chemin de Fer à Vapeur des 3 Vallées (CFV3V). Easily the largest museum set-up in the Benelux countries. Originally operating over the closed line to Treignes, operations have recently been extended in the opposite direction from Mariembourg to Chimay and even further still to Momignies on certain days. (A further 13 km). The main depot is at Mariembourg in the old roundhouse there, but locos are often away for repair nearer the main centres of population. Recent visits have found CFV3V locos at Haine St. Pierre, Ronet and Salzinnes Works! The line operates weekends and holidays April–September with trains also running Tuesdays–Thursdays in July and August, not necessarily over the whole line. The last weekend in September is usually festival weekend.
21 steam, 4 diesels, 8 railcars.

Mechelen

It is understood a new museum has opened in part of the railway workshops.The replica locomotives formerly stored at Tournai are believed to be here. Reported as open on the last Saturday of the month.

Rebecq–Rognon. 3 km.

600 mm. Rail Rebecq–Rognon. RRR. Located near Tubize from where there is an infrequent bus service or a 45 minute walk. The line serves a pleasant area along the River Senne. Operates Sundays and holidays May–September.
3 steam, 3 diesel.

Schepdaal.

600, 1000, 1435 mm. Association pour le Musée du Tramway. AMUTRA. A large collection of tramway equipment which includes 3 steam locos. Located 13 km from Brussels, access is by bus N Brussels–Ninove. Open Sundays and holidays Easter–October and also Saturdays in July and August, 14.00–18.00.
3 steam, 3 diesel railcars, many trams.

Sprimont.

600 mm. Chemin de Fer de Sprimont. CFS. This line is built on the trackbed of a former SNCV line and uses old mining diesel locos. Operates first and third Sundays of the month May–September from 14.00. Festival on the last weekend of August. Public transport access is believed to be by bus from Poulseur–Trooz. Nearest main towns are Liège and Spa.
6 diesels.

Trembleur. 5 km.

1000 mm. Complex Touristique Li Trembleu. A tourist centre has been opened in this old mining town. Besides the operating railway, visits to the mine are also possible. Access is by bus STIL 67 from Liège. Complete stock is not known. Open daily May–September.

Zolder–Genebos. 10 km.

Toeristische Trein Zolder. TTZ. Services operate over former colliery lines and it is possible to visit the pit. Operates Sundays and holidays late June to early September. Access is Zolder station.
6 steam, 1 diesel, 1 DMU, plus 3 narrow gauge locos.

NETHERLANDS

Amsterdam

Electrische Museumtramlijn. Amsterdam has a large collection of trams and trailers and operates over the city network. Most of the stock is at the Karperweg depot but some may also be found at Havenstraat depot.
31 trams, 20 trailers, 2 electric locomotives, 2 diesel locomotives and numerous other tramway type equipment.

Apeldoorn–Dieren. 22 km.

Veluwsche Stoomtrein Maatschappij. VSM. Operates over a closed NS branch line on which the depot is located at Beekbergen. Main operating days are Mondays–Fridays and Sundays in July and August. Uses mostly former DB locos.
4 steam, 2 diesel, 1 DMU.

Barger Compascuum. 1 km.

700 mm. Eerste Grentse Vereniging van Stoomliefhebbers EDS. This operation is in the National Veenpark and a short train journey is made through the old village. This Moorland Park is open every day from April to October. The train rides are included in the entrance fee.
2 steam, 10 diesels.

Beverwijk. 9 km.

Hoogovens Excursietrein. This is a steam hauled excursion train around the vast steelworks covering 18 kms return journey which takes some 2 hours. Photo stops are made on the journey. Departure is from Beverwijk station at 10.45 on the last Sunday of the month May to September.
2 steam. (Plus all the industrial diesels in the steelworks!).

Erica.

700 mm. Industrieel Smalspoormuseum te Erica. A rather new museum about which little is known except the rolling stock.
22 diesels, 3 electrics.

Goes–Oudelande. 16 km.

Stoomtram Goes–Borsele. SGB. This line uses a former NS freight line. The depot is located at Goes station and is in fact the old NS depot. In July and August the line operates daily except Saturdays. Early and late season operations are normally Sundays only although the third weekend in September is usually an enthusiasts weekend. Departures in the main season are at 14.05 with an evening wine and dine train leaving at 18.30 on Wednesday. The round trip takes nearly 3 hours.
4 steam, 1 diesel.

Haaksbergen–Boekelo. 7 km.

Museum Buurt Spoorweg. MBS. The depot is located at Haaksbergen which is 9 km from Enschede and can be accessed by GSM buses 20, 21, 22, 27 from there or TET 53 bus from Hengelo. Operates on Wednesdays and Sundays in July and August.
5 steam, 6 diesels, 3 DMUs.

Den Haag.

Haagse Tram Museum. HTM. Museum trams operate on certain Saturdays or Sundays in the summer.

Hoorn–Medemblik. 20 km.

Stoomtram Hoorn–Medemblik. SHM. This line is well established and a visit is recommended. The depot is adjacent to Hoorn NS station. NS day trip tickets can be purchased if a railrover is not being used. It is possible to do a train–boat–train circular journey. During July/August operation is daily whilst at other times between May and September there are no trains on Mondays.
9 steam, 5 diesels, 1 DMU.

Kaatsheuvel.1.5 km.

600 mm. Efteling Stoomtrein Maatschaapij. This is a steam operation in a vast amusement park close to Tilburg. The park is open and trains run each day April to October between 10.00–18.00.
? steam, ? diesel.

Katwijk. 4 km.

700 mm. Nederlandse Smalspoorweg Stichting. NSS. An interesting line set amongst the sand dunes alongside the coast near Leiden. The area serves as a reservoir for Leiden and only a few operating days take place each year as the area is also a conservation area. The society has an ever growing fleet of steam and diesel locos, so much so that some are stored under cover at the SGB depot in Goes. The NSS has plans to build a proper narrow gauge museum at nearby Valkenburg. 1988 operations were on Saturdays from mid July to mid September. Access by bus from Leiden (NZH 30, 31, 41), Den Haag CS (NZH 90, 91) or Haarlem (NZH 90, 91).
13 steam, 42 diesels. (Some stock is 900 mm).

Ouddorp.

1067 mm. Stichting Rijdend Tram Museum. RTM. This line was previously located at Hellevoetsluis and was in the process of moving to this new site early in 1989. The depot is believed to be located at Brouwersdam near Ouddorp but no details of their commencement operations there are known nor the length of the new line.
3 steam, 3 DMUs, 1 diesel.

Roosendaal

Stichting tot Behoud van Af te voeren NS materieel. STIBANS. This society is currently occupying the old NS depot at Roosendaal. Besides its own stock it also looks after certain items from the Utrecht collection.
1 electric, 1 EMU.

Rotterdam

Rotterdamse Elektrische Tramwegmaatschappij. RET. During the summer the Stichting Brabant Rail have been running their steam tram from Centraal station to Willemsplein usually hourly between 10.00 and 15.00. Additionally Tramweg Stichting has preserved several Rotterdam trams and these may be used on sightseeing trips almost at anytime.

Rotterdam

Stoom Stichting Nederland. SSN. This society's depot is located at Spaanse Polder, Giessenweg. Take tram RET 1/6/8 from Rotterdam or even the metro to Marconiplein or alternatively bus RET 38 from Schiedam to Franselaan. The SSN does not have a museum line but is able to run its locos several times each year on excursion trains over the NS. Most of the locos are former DB ones.
7 steam, 2 diesel.

Utrecht.

Nederlands Spoorweg Museum, Maliebaanstation, Jahan van Oldenbarneveldtlaan 6. During recent years the collection here has grown and in readiness for the NS 150 celebrations certain renovations and extensions have taken place.The museum is now fully open again and the hours are expected to be 10.00–17.00 (Tuesday–Saturday) and 13.00–17.00 (Sundays). Bus No. 3 from Utrecht CS should be used to Rubenslaan.
14 steam, 4 diesel, 2 EMUs, 2 DMUs and 8 trams and other equipment.

LUXEMBOURG.

Luxembourg.

CFL Depot. Staff here have kept two diesel locos as museum pieces and are restoring 2–10–0 5519 at Pétange wagon works. It is very likely that this loco will get a main line ticket.
1 steam, 2 diesels.

Luxembourg.

CFL Depot. Groupement des Amis du Rail. GAR. This society is also based at the CFL Depot and has a former ÖBB Class 52 2–10–0 and a CFL diesel railcar. The railcar is nearly ready to run again.

Rodange. 5 km.

Association des Musée et Tourisme Ferroviaires. AMTF. The depot for this line is at Fond de Gras. Access to the line is Rodange station and a mile walk to Bois de Rodange where the museum line station is located. Operations are on Sundays and holidays May to September. Usually three trips during the afternoon are made by the steam train and a DMU is also used if another steam loco is not available for the second train or traffic is light.
7 steam, 2 diesel, 1 DMU.

'Wadloper' (Mud-flat runner) logo on the side of an NS DMU. *Peter Fox*

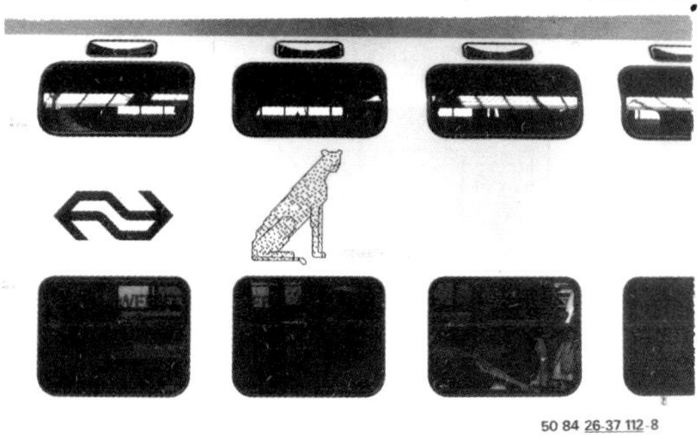

The NS double-deck driving trailers are named after endangered species. This photograph shows the side of 50 84 26-37 112-8 'Cheeta' *Peter Fox*

Thanks are due to the readers of our first edition who pointed out that although the official language of Luxembourg is French, most of the population speak Letzeburgisch, a low dialect of German. Here is another nameplate with inscriptions in this language. *Alexander Foot*

Further copies of this publication can be obtained from the address on the title page, price £5.95 plus 10% postage (UK) or 20% (overseas). Payment must be by UK cheque or P.O. or by Eurocheque or International Money Order in pounds sterling. We do not accept US $ cheques.
We also publish rolling stock guides for the UK, France, West Germany and Austria and a book on Switzerland is being written.